Radiation Safety Problems in the Caspian Region

T0180618

NATO Science Series

A Series presenting the results of scientific meetings supported under the NATO Science Programme.

The Series is published by IOS Press, Amsterdam, and Kluwer Academic Publishers in conjunction with the NATO Scientific Affairs Division

Sub-Series

I. **Life and Behavioural Sciences**	IOS Press
II. **Mathematics, Physics and Chemistry**	Kluwer Academic Publishers
III. **Computer and Systems Science**	IOS Press
IV. **Earth and Environmental Sciences**	Kluwer Academic Publishers
V. **Science and Technology Policy**	IOS Press

The NATO Science Series continues the series of books published formerly as the NATO ASI Series.

The NATO Science Programme offers support for collaboration in civil science between scientists of countries of the Euro-Atlantic Partnership Council. The types of scientific meeting generally supported are "Advanced Study Institutes" and "Advanced Research Workshops", although other types of meeting are supported from time to time. The NATO Science Series collects together the results of these meetings. The meetings are co-organised bij scientists from NATO countries and scientists from NATO's Partner countries – countries of the CIS and Central and Eastern Europe.

Advanced Study Institutes are high-level tutorial courses offering in-depth study of latest advances in a field.
Advanced Research Workshops are expert meetings aimed at critical assessment of a field, and identification of directions for future action.

As a consequence of the restructuring of the NATO Science Programme in 1999, the NATO Science Series has been re-organised and there are currently five sub-series as noted above. Please consult the following web sites for information on previous volumes published in the Series, as well as details of earlier sub-series.

http://www.nato.int/science
http://www.wkap.nl
http://www.iospress.nl
http://www.wtv-books.de/nato-pco.htm

Series IV: Earth and Environmental Sciences – Vol. 41

Radiation Safety Problems in the Caspian Region

edited by

Mohammed K. Zaidi
US Department of Energy,
Idaho Operations Office, Idaho Falls, Idaho, U.S.A.

and

Islam Mustafaev
Institute of Radiation Problems,
Azerbaijan National Academy of Sciences, Baku, Azerbaijan

Kluwer Academic Publishers

Dordrecht / Boston / London

Published in cooperation with NATO Scientific Affairs Division

Proceedings of the NATO Advanced Research Workshop on
Radiation Safety Problems in the Caspian Region
Baku, Azerbaijan
11–14 September 2003

A C.I.P. Catalogue record for this book is available from the Library of Congress.

ISBN 1-4020-2377-4 (PB)
ISBN 1-4020-2376-6 (HB)
ISBN 1-4020-2378-2 (e-book)

Published by Kluwer Academic Publishers,
P.O. Box 17, 3300 AA Dordrecht, The Netherlands.

Sold and distributed in North, Central and South America
by Kluwer Academic Publishers,
101 Philip Drive, Norwell, MA 02061, U.S.A.

In all other countries, sold and distributed
by Kluwer Academic Publishers,
P.O. Box 322, 3300 AH Dordrecht, The Netherlands.

Printed on acid-free paper

TABLE OF CONTENTS

List of Participant ix

List of contributor addresses xiii

Preface xxi

SESSION 1: *RADIATION-ENVIRONMENTAL MONITORING IN THE CASPIAN REGION.*

Remote Radiation Environment Monitoring 1
A.M. Pashayev, A.Sh. Mehdiev and A.A. Bayramov

Monitoring of the Radioecological Situation in Marine and Coastal
Environment of Georgia 5
M. Avtandilashvili, D. Baratashvili, R. Dunker, N. Mazmanidi, S. Pagava, Z. Robakidze,
V. Rusetski and G. Togonidze

Ecologic Situation of the Aral Sea Region 13
E. Oteniyazov

Distribution of Heavy Metals in the Soil of Industrial Zones of Apsheron Peninsula   17
L.A. Aliyev

Structural Peculiarities of Borosilicates and its Correlation with
Radiation-Catalytic Activity 23
A.M. Gasanov, E.A. Samedov and S.Z. Melikova.

Characteristics of Broken Grounds at the Former Azgir Nuclear Test Site 29
E.Z. Akhmetov, Zh.I. Adymov, V. Dzhezairov-Kakhramanov and A.S. Yermatov

SESSION 2: *RADIATION-ENVIRONMENTAL MONITORING METHODS - GENERAL ISSUES.*

Early Warning Environmental Radiation Monitoring System (RESA) 33
N. Küçükarslan, A. Erdoğan, A. Güven and Y. Gülay

Uranium Production as a Byproduct from Yarimca Phosphoric Acid Plant 43
G. Önal and S. Atak

Biotesting of Radiation Pollutions Genotoxicity with the Plants Bioassays 51
N. Kutsokon, N. Rashydov, V. Berezhna and D. Grodzinsky

Structure of Met30-Ser40 Segment in N-terminus of Human Tyrosine Hydroxylase Type-1 57
Irada N. Alieva, Narmina Mustafayeva, and Dshavanchir Aliev

SESSION 3: *NATURAL AND ARTIFICIAL SOURCES OF RADIOACTIVITY IN THE CASPIAN REGION.*

Ecological Situation at "Koshkar-Ata" Nuclear Test Site 69
K.K. Kadyrzhanov, K.A. Kuterbekov, S.N. Lukashenko, V.N. Gluschenko and M.M. Burkitbaev

Problem of Risk Modeling: Influence of Uranium Storage on Environment 79
A.K. Tynybekov

Radioactive Minerals and Nuclear Fuels in Turkey 85
Dündar Renda

Solvent Extraction of Uranium from Wet Process Phosphoric Acids 89
Seref Girgin, Ayhan Ali Sirkeci and Neset Acarkan

Radioactivity of Lakes in the Urbanized Territories 97
V.A. Mammadov

Natural Radionuclides in Soil-plants in Sheki-Zakatala Zone of Azerbaijan 103
A.A. Garibov, I.A. Abbasova. M.A. Abdullayev and Ch.S. Aliyev

SESSION 4: *RADIODIAGNOSTICS AND RADIOTHERAPY.*

**Prevention of Accidental Exposures to Radiodiagnostics and
Radiotherapy Patients – Radiation Safety Aspects** 107
Mohammed K. Zaidi and Thomas F. Gesell

New Approach to Use of Polyene Antibiotics 121
V.Kh. Ibragimova, I.N. Alieva and D.I. Aliev

SESSION 5: *RADIATION SAFETY PROBLEMS IN OIL INDUSTRY*

**Radiological Impact on Man and the Environment from the Oil and Gas Industry:
Risk Assessment for the Critical Group** 129
F. Steinhäusler

Radiation–Thermal Purification of Waster Water from Oil Pollution 135
I. Mustafaev, N. Guliyeva, S. Aliyev and I. Mamedyarova.

Radiation–Thermal Refining of Organic Parts of Oil-Bituminous Rocks 141
I. Mustafav, L. Jabbarova, N. Guliyeva and K. Yagubov

Radio-ecological State of Absheron Oil and Gas Extracting Fields 147
Reavn. N. Mehdiyeva, Hokman.M. Mahmudov and M.F. Gaffarov.

**Influence of Technological Cycles of Natural Gas Treatment on Radioactive
Radon Content in its Composition** 151
A.A. Garibov, G.F. Miralamov, R.Ch. Mamedov and G.Z. Velibekova.

Marine Gamma Survey of Seabed of Caspian Sea 157
Z.T. Bayramov and A.M. Gasanov

SESSION 6: *EXPORT CONTROL OF NUCLEAR MATERIALS*

**Outlook to Nonproliferation Activities in the World and Cooperation in Peaceful
Uses of Nuclear Energy among Turkey, Caucasian and Central Asian Countries** 161
Nevzat Birsen

Emerging Nuclear Security Issues for Transit Countries 165
I.A. Gabulov

Solution of Questions of Non-Proliferation in Georgia 169
Anatol Gorgoshidze

Nuclear Terrorism and Protection of Ecology 173
Zaur Ahmad-zada and Dzhavanshir Aliev

SESSION 7: *NEW METHODS AND TECHNICS OF DOSIMETRY OF IONIZING RADIATION*

Thermoluminescence Personal and Medical Dosimetry 177
Mária Ranogajec-Komor.

Aqueous-Phase Chemical Reactions in the Atmosphere 191
A.N. Yermakov, I.K. Larin and A.A. Ugarov

**Measurement of Natural Radioactivity in Phosphogypsum by High Resolution
Gamma Ray Spectroscopy** 197
*H. Yücel, H. Demirel, A. Parmaksız, H. Karadeniz, İ. Türk Çakır, B. Çetiner,
A. Zararsız, M. Kaplan, S. Özgür, H. Kızılal, M.B. Halitligil and İ. Tükenmez*

Electret Polymer Materials for Dosimetry of ▫-irradiation 205
A.M. Magerramov, M.K. Kerimov and E.M. Hamidov

SESSION 8: ENVIRONMENTAL *IMPACT OF ELECTROMAGNETIC FIELD. RADAR STATIONS.*

Environmental Impact of Electromagnetic Fields 211
Edwin Mantiply

Ecological Problems of Power Engineering – Electromagnetic Compatibility and Electromagnetic Safety 217
M.S. Qasimzade, F.S. Aydayev and V.M. Salahov

SESSION 9: *NGO-ACTIVITY IN THE CASPIAN REGION ON RADIATION SAFETY.*

The Professional-Oriented Regional Radioecological Collaboration of Southern Caucasian States 225
M. Avtandilashvili, S. Pagava, Z. Robakidze and V. Rusetski

Public Opinion and its Influence on Prospects of Development of Nuclear Power 231
E.A. Rudneva, A.V. Rudnev and N.P. Tarasova

Analysis of Radio-ecological Situation in Azerbaijan 241
L.A. Aliyev

Subject Index 247

NATO ARW 2003 BAKU - LIST OF PARTICIPANTS

	Name	Country, Organization	Status	Address
		AZERBAIJAN HOST COMMITTEE		
1.	Mahmud Kerimov	Azerbaijan, Academy of Sciences (ANAS)	Chairman of the Organizing Committee	President@lan.ab.az;
2	Garibov Adil	Institute of Radiation Problems (IRP), ANAS	Key Speaker	Ibrahim_Gabulov@yahoo.com
3	Gabulov Ibrahim	Institute of Radiation Problems (ANAS)	Speaker	Ibrahim_gabulov@yahoo.com
4	Mustafaev Islam	Institute of Radiation Problems (ANAS)	Co-Director, Key Speaker	imustafaev@iatp.az
		AUSTRIA		
5	Steinhausler Friedrich	Institute of Physics and Biophysics University of Salzburg, Austria.	Key speaker	physik@sbg.ac.at friedrich.steinhaeusler@sbg.ac.at
		CROATIA		
6	Ranogajec-Komor Maria	Croatia, Ruđer Bošković Institute, Zagreb, Croatia.	Key Speaker	marika@irb.hr
		GEORGIA		
7	Rusetski Vladimir	Georgia, Tbilisi State University	Key Speaker	spagava@access.sanet.ge
8	Gorgoshidze Anatol	Georgia, Border Guard Service	Speaker	anatol_gorgoshidze@hotmail.com,
9	Pagava Samson	Georgia, NGO "Radiolog-21"	Speaker	spagava@access.sanet.ge
		KAZAKHSTAN		
10	Burkitbayev Mukash	Kazakhstan, al-Farabi Kazakh State University	Key Speaker	nburkit@nursat.kz
11	Brodskaya Yulia	Kazakhstan, Resource Center for Ecological NGOs	Speaker	rc-ecoforum@carec.kz
12	Kahramanov Veysel	Kazakhstan, Institute of Nuclear Physics	Speaker	office@tandem-translations.com

13	Kairat.A. Kuterbekov	Kazakhstan, Institute of Nuclear Physics	Speaker	kuterbekov@inp.kz

KYRGYZSTAN

14	Tynybekov Azamat	Kyrgyzstan, NGO ISC	Speaker	isc@freenet.kg
15	Charskiy Vyacheslav	Kyrgyzstan, NGO For Civil Society	Speaker	root@agat.freenet.bishkek.su

RUSSIA

16	Yermakov Alexander	Russia, Russian Academy Sciences	Speaker	polclouds@mtu-net.ru
17	Rudneva Evgeniya	Russia, Youth Center - The Creative Person Development.	Speaker	rudalex@mail.ru
18	Lagutov Vladimir	Russia, NGO "Green Don"	Speaker	zedon@novoch.ru

TURKEY

19	Birsen Nevzat	Turkey, TAEK	Key speaker	n.birsen@taek.gov.tr
20	Yucel Haluk	Turkey, TAEK	Speaker	haluk.yucel@taek.gov.tr
21	Kuchukarslan Necati	Turkey, TAEK	Key speaker	necati@nukleer.gov.tr
22	Onal Guven	Turkey, ITU	Speaker	onalg@itu.edu.tr
23	Sirkeci Ayhan	Turkey, ITU	Speaker	sirkecia@itu.edu.tr
24	Renda Dundar	Turkey, Miners Association	Speaker	drenda@superonline.com

TURKMENISTAN

25	Ishankuliyev Dovlatyar	Turkmenistan, State University	Speaker	geldy_ishankuliev@hotmail.com

UKRAINE

26	Voitsekhovski Oleg	Ukraine, Ministry of Environmental Protection	Speaker	voitsekh@voi.vedos.kiev.ua
27	Rashydov Namik	Ukraine, Ukrainian Academy Sc	Speaker	nrashydov@yahoo.com

UNITED STATES OF AMERICA (USA)

28	Zaidi Mohammed	US Department of Energy	Co-Director	zaidimk@id.doe.gov

				Key Speaker	
29	Mantiply Ed	US Federal Communication Commission	Key speaker	EMANTIPL@fcc.gov

UZBEKISTAN

30	Oteniyazov Esbosin	Uzbekistan NGO "Aral Defense Committee"	Speaker	Bakhytn@yandex.ru

AZERBAIJAN

31	Gurbanov Muslim	Azer, NGO "Ecoil"	Speaker	mgurbanov@hotmail.com
32	Mammedov Vagif	Azer, Institute of Geology	Speaker	vmamedov@hotmail.com
33	Ojagov Habib	NGO "Fovgal"	Speaker	
34	Aliyev Chingiz	Azer, Inst of Geology	Speaker	
35	Suleymanov Bahruz	Comp "Azerecolab"	Speaker	suleymanov@azecolab.com
36	Samadova Ulviyya	IRP, ANAS	Speaker	U_samadova@yahoo.com
37	Bayramov Azad	Azer Institute of Physics	Speaker	bayramov_azad@mail.ru
38	Jafarov Elimkhan	IRP, ANAS	Speaker	
39	Bayramov Zakir	IRP, ANAS	Speaker	
40	Velibekova Gulya	IRP, ANAS	Participant	
41	Mehdiyeva Revan	IRP, ANAS	Participant	
42	Mammedyarova Irina	IRP, ANAS	Participant	Imamedyarova@yahoo.com
43	Madatov Rahim	IRP, ANAS	Participant	
44	Abbasov Shirin	IRP, ANAS	Participant	
45	Mahmudov Hokmen	IRP, ANAS	Participant	
46	Magerramov Arif	IRP, ANAS	Participant	
47	Guliyeva Nigar	IRP, ANAS		
48	Yagubov Kamal	IRP, ANAS	Participant	
49	Aliyev Selimxan	IRP, ANAS	Participant	
50	Gasanov Aflatun	Azer AeroSpace Agency	Participant	
51	Aliyev Lahuti	Azerbaijan Ministry of Ecology and N.Resources	Participant	monitoring@mmd.baku.az
52	Aliyev Ayaz	Azerbaijan Ministry of Ecology and N. Resources	Participant	
53	Yuzbashova Sevil	NGO "For Clean	Participant	

		Caspian"		
54	Rzayeva Nailya	Baku Oxford School	Participant	Gunar_educate@bak.net.az
55	Aliyev Anar	Post –graduate student	Participant	
56	Aliyev Sabuhi	Student	Participant	wide@box.az
57	Salahov Vilayet	Graduate Student	Participant	
58	Heydarova Sevinc	Azerbaijan	Mass-media	
59	Mustafaev Ilgar	Azer. NGO Kraevedeniye	Video-Photo Operator	imustafaev@mail.ru

NATO ARW 2003 – ADDRESSES OF CONTRIBUTERS

1. Reference number - 979498

2. Title of ARW - " RADIATION SAFETY PROBLEMS IN THE CASPIAN REGION"

3. **Director from NATO country:**

Mohammed K. ZAIDI, Physicist,
U.S. Department of Energy, Idaho Operations Office
Radiological and Environmental Sciences Laboratory
1955, Fremont Avenue
Idaho Falls, ID 83401-4149. USA.

Tel: 208-526-2132,
Fax: 208-526-2548,
Res: 208-234-1130.
zaidimk@id.doe.gov

4. **Director from Partner country:**

Dr. Islam MUSTAFAEV
Chairman Ecological Society "Ruzgar"
Institute of Radiation Problems
Azerbaijan National Academy of Sciences
31-a H. Javid Avenue, 370143
Baku, AZERBAIJAN.

Tel: 99412-39-4113, Mobile: 99450-3207816
Fax: 99412-743004, 99412-398318.
imustafaev@iatp.baku.az; imustafaev@iatp.aznet.org
www.aznet.org

5. **NATO Executive:**

Dr. Alain Jubier, Program Director,
Environmental & Earth Sciences and Technology Programs
NATO Headquarters, Bd. Leopold III
B-1110 BRUSSELS
BELGIUM.

Tel: 32-2-707.5041
Fax: 32-2-707.5057
E-mail: science.est@hq.nato.int

6. PARTICIPANTS/CONTRIBUTERS:

Edwin Mantiply, Physical Scientist
Federal Communications Commission
Office of Engineering and Technology
Room 7-A201, 445 12th Street, SW
Washington, DC. 20554, USA.

Tel: 202-418-2423
Fax: 202-418-1918
emantipl@fcc.gov

Prof. Friedrich Steinhausler
Institute of Physics and Biophysics
Hellbrunnerstr. 34, A-5020 Salzburg
AUSTRIA.

email - direct: friedrich.steinhausler@sbg.ac.at
email - office: physik@sbg.ac.at
phone or fax: +43-(0)662-8044-5701
fax (office):   +43-(0)662-8044-150
mobile phone: +43-(0)676-304 8256

Dr. Adil Garibov
Institute of Radiation Problems
Azerbaijan National Academy of Sciences
31-a H. Javid Ave., 370143
Baku, AZERBAIJAN.

Dr. Mahmud Kerimov, President
Azerbaijan National Academy of Sciences
31-a H. Javid Avenue, 370143
Baku, AZERBAIJAN.

Tel: 99412-393391
Fax: 99412-398318
rad@dcacs.ab.az

Dr. Samson Pagava, Head
Radiocarbon and Low-Level Counting Section
I.Javakhishvili Tbilisi State University
Physics Faculty, 3, I. Chavchavadze
Tbilisi 380028, GEORGIA

spagava@access.sanet.ge

Azamat Kalyevich Tynybekov
Room 38, Second Floor
40, Manas Street, Bishkek,
Kyrgyz Republic.

isc@freenet.kg
www.isc.freenet.kg
996-312-220557, Fax – 996-312-481-806

Dr. Alexandr Yermakov, Head of Laboratory
Leninsky pr-t, 38, Building 2
Institute of Energy Problems of Chemical Physics
Russain Academy of Sciences
38 Leninsky Pr, 117337
Moscow, RUSSIA.
Tel: 095-422-5005
ayermakov@chph.ras.ru

Dr. Ayhan A. Sirkeci
Istanbul Technical University
Mining Faculty, Dept. of Mining Engr.
34469 Maslak, Istanbul, Turkey.

Fax. +90-212-285 61 28
Tel. +90-212-285 30 11
Mobile. +90-533-421 18 02.
sirkecia@itu.edu.tr

Dr. Nevzat Birsen
Turkish Atomic Energy Agency
Ankara, TURKEY
tudnaem@taek.gov.tr

Oleg V. Voitsekhovitch, Ph.D. Head
Environmental Radiation Monitoring Department
Ukrainian Hydrometeoroloigical Institute
37, Prospect Nauky
Kyiv, Ukraine, 03028
voitsekh@voi.vedos.kiev.ua
Fax 380-44 2651130, Phone: 380442658633

Maria Ranogajec-Komor, Ph.D.
Radiation Chemistry and Dosimetry Laboratory
Ruder Boskovic Institute, P. O. Box 180
1002 Zagreb, Bijenicka 54, CROATIA.

Marika@irb.hr

Dr. Anatol Gorgoshidze
State Department of Border Defense,
Head of Office of Development of Border Guards
12 Kandelaki Street, Tbilisi 380060, GEORGIA.

Anatol_gorgoshidze@hotmail.com

V. Rusetski
Radiocarbon and Low-Level Counting Section
Nuclear Research Laboratory, Physics Faculty,
I. Javakhishvili Tbilisi State University
Chavchavadze Avenue 3, Tbilisi 380028, GEORGIA.

Mukhambetcali Burkitbayev, Head
Chair of Inorganic Chemistry
Al-Farabi Kazakh National University
95 Karasai batyr Street
480012, Alamaty, KAZAKHSTAN.

mburkit@nusrat.kz

Kairat Kuterbekov
Institute of Nuclear Physics, National Nuclear Centre
Office of the Scientific Secretary
Semipalatinsk, Alamaty 480082, KAZAKHSTAN.

kuterbekov@inp.kz , FAX: +7 3272 546517, Ph +7 3272 545652

V. Dzhezairov-Kakhramanov
Institute of Nuclear Physics, National Nuclear Centre
Office of the Scientific Secretary
Semipalatinsk, Alamaty 480082, KAZAKHSTAN.

Yuliya Brodskay, Manager
Ecological Campaigns, Anti-Nuclear Movement
Mkr Orbita-1, house 40
Almaty 480043, Kazakhstan

Yuliya_brodskaya@newmail.com
www.rcecoforum.narod.ru
8-333-237-0200.

Dr. Vladimir Komlev, Senior Scientist
Kola Science Centre,
Russian Academy of Sciences
Moscow, RUSSIA.

komlev@goi.kolasc.net.ru

Evgeniya A. Rudneva,
Noncommercial Organization, Youth Centre
The Creative Person Development
Str. Gruzinsky val 28/45, Moscow, RUSSIA.

Evgeniya Rudneva - rudalex@mail.ru, rudneva@newmail.ru

V. Lagutov,
NGO Regional Ecological Movement "Green Home"
Dachnaya Street 1,2, Novocherkassk 346408, RUSSIA.

zedon@novoch.ru

Dündar Renda
Istiklal Cad. Tunca Apt. No. 471/ 1-1
Tunel, Istanbul, TURKEY.

drenda@superonline.com, turkiyemaden@ixir.com

Prof. Dr. Güven Önal
Istanbul Technical University, Mining Faculty
Ayazaga, 80626, Istanbul, TURKEY.

onalg@itu.edu.tr

Dr. Necati Kucukarslan
Kucuk Cekmeci, Istanbul, Turkey.

necati@nukleer.gov.tr

Dr. Haluk. Yücel,
Turkish Atomic Energy Authority (TAEA)
Ankara Nuclear Research and Training Center
06100 Beşevler, Ankara, TURKEY.

Haluk.yucel@taek.gov.tr

Dr. Ishankuliev Dovletyar
Natural and Artificial Radiation in Turkmenistan
Center for Physical and Mathematical Research
Turkmen State University, Ashgabat, TURKMENISTAN.

Geldy_ishankuliev@hotmail.com

Dr. N. Kutsokon, Institute of Cell Biology
and Genetic Engineering,
National Academy of Sciences of Ukraine
Kiev, UKRAINE.

nrashydov@yahoo.com

Dr. Ibrahim Gabulov
Institute of Radiation Problems
Azerbaijan National Academy of Sciences
H. Javid Avenue, 31A, 371143
Baku, AZERBAIJAN.

Ibrahim_gabulov@yahoo.com

V. Kh. Ibragimova, Institute of Radiation Problems
Azerbaijan National Academy of Sciences
H. Javid Avenue, 31A, 371143
Baku, AZERBAIJAN.

iradanur@mail.az

Zaur Ahmad-zada
Institute of Human Rights
National Academy of Sciences of Azerbaijan
H. Javid Avenue, 31A, 371143
Baku, AZERBAIJAN.

Bianco_nero@mail.ru

Dr. Azad. A. Bayramov
Azerbaijan National Aviation Academy
Bina Aeroport 370057
Baku, Azerbaijan.

bayramov_azad@mail.ru
Tel: 97-26-20, Fax: 97-28-29, E-MAIL:

Dr. V. A. Mammadov
Geology Institute,
Azerbaijan National Academy of Sciences
H. Javid Avenue, 31A, 371143
Baku, AZERBAIJAN.

radiometry@gia.ab.az

M. S. Qasimzade
Azerbaijan Scientific–Research Institute
Power Engineering and Power Designing
Baku, Azerbaijan.

vilayet79@yahoo.com

Dr. Irada Alieva
Laboratory of Molecular Biophysics
Department of Physics, Baku State University
Z.Khalilov str.,23 AZ1073/1, Baku-Azerbaijan.

Phone:(+99412) 321 978, iradanur@mail.az or iradanur@box.az

Dr. Etib.A. Samedov
Azerbaijan National Aerospace Agency
Institute of Radiation Problems
National Academy of Sciences
159 Azadlig Avenue, AZ1106, Baku, AZERBAIJAN.

sevincmelikova@yahoo.com

Dr. Zakir T. Bayramov
National Academy of Sciences
159 Azadlig Avenue, AZ 1106, Baku, Azerbaijan.

bayramov_z@yahoo.com

Dr. Lahuti A. Aliyev.
National Academy of Sciences
Ministry of Ecology and Natural Resources
Baku, AZERBAIJAN.

Dr. R. N. Mehdiyeva
Institute of Radiation Problems,
National Academy of Sciences,
159 Azadlig Avenue, AZ1106
Baku, AZERBAIJAN.

hokman@rambler.ru

Dr. A. M. Magerramov
Institute of Radiation Problems
National Academy of Sciences,
H. Javid av. 31 A, Baku, AZ 1143, Azerbaijan.

Arifm50@rambler.ru, elsevar1966@mail.ru

PREFACE:

This North Atlantic Treaty Organization (NATO) Advanced Research Workshop (ARW) was devoted to the Radiation Safety Problems in the Caspian Region. Altogether more than 60 papers were presented; only 37 research papers could be collected and put in this proceedings manual for ARW.

This ARW was held at Mardakan Dendrary Park Convention Center, Baku, Azerbaijan during September 11-14, 2003. Financial support came from the NATO Scientific Committee and also from the East-East Foundation. The Azerbaijan National Academy of Sciences (ANAS) staff provided local logistical support.

Seventy-six participant originated in twelve countries, i.e. Austria, Azerbaijan, Croatia, Georgia, Kazakhstan, Kyrgyzstan, Russia, Turkey, Turkmenistan, Ukraine, United States of America and Uzbekistan.

The Organizing Committee of this ARW consisted of Dr. M. Kerimov, President of ANAS, and Dr. A. Garibov (ANAS), N. Birsen (Turkey), A. Yermakov (Russia), M. Burkitbayev (Kazakhstan) and S. Pagava (Georgia).

Dr. Islam Mustafaev chaired the Executive Organizing Committee with the following membership: G. Mehdiyeva, I. Mamedyarova, G. Velibeyova and I. Gabulov. K. Mustafaev served as the technical secretary for the committee. Irina Mamedyarova was the publishing editor and graphic designer for the ARW program and Book of Abstracts.

All of the above noted have to be thanked for their services, specially the co-director, Dr. Mustafaev. He personally welcomed every participant at the airport even though some flights arrived very late or very early in the morning. I am personally thankful to all the participants for taking their time to come and participate. Furthermore, I want to express my gratitude to those who had submitted their papers for publication.

I am thankful to RESL Director, Dr. R.D. Carlson and RESL staff for helping me to do this job, Prof. Dr. Thomas F. Gesell for his support during the preparatory period of this ARW, and Dr. Kerimov for his thoughtfulness to help me organize this ARW. Dr. Allan Jubier, Director, NATO Scientific Programs and his staff were of great help during the proposal submission, revision of the proposal, its award, the final reports, financial report and the submission of this manuscript. I am thankful to my wife, Shahnaz, for helping me and giving me moral support during this painful job of editing this proceedings manual.

REMOTE RADIATION ENVIRONMENT MONITORING

A. M. PASHAYEV, A. Sh. MEHDIEV and A. A. BAYRAMOV

National Aviation Academy, Airport Bina, Baku, Azerbaijan

Corresponding author: bayramov_azad@mail.ru

ABSTRACT:

The environment radiation monitoring system was developed. Measurements with the help of "EKOMON" fixed stations, consisting of appropriate sets of registration, the analysis and transfer of received data on an off-wire communication circuit and an autonomous supply set being planned to monitor ecological environmental factors: a level of an electromagnetic background, density dangerous and noxious gases, temperatures, pressure, atmosphere and wind speed. It will allow more precisely and to forecast development of an ecological situation not only in Azerbaijan, but also in whole of the neighboring states.

Keywords: Radiation safety, environment, automated monitoring, emergencies and ecological catastrophes.

INTRODUCTION:

Azerbaijan Republic doesn't have nuclear apparatuses, nuclear reactor, nuclear energy systems and nuclear materials technologies. But our country has frontier with the countries, which have nuclear technologies. Before the accident at the Chernobyl Atomic Power plant the greatest concentrations of ^{90}Sr and ^{137}Cs were 1-30 Bk/kg. After the accident, the concentrations of these radionuclides in the earth levels were increased by 2-3 times. At the same time the quantity of radionuclides falling from atmosphere were also changed accordingly.

Evidently the accident in the nuclear reactor, the waste will affect the air, water basins, and the ground surface of Azerbaijan. It will also affect on the mass of radionuclides. The safe development of nuclear technology of our neighbor countries is important for the safety of the people of Azerbaijan. Next new developments of our neighbors are important in radiation safety of our Republic:

1. Armenian Nuclear Electric Power Station and reactions operating there.
2. WER type, 316 MW reactor - Atomic power plant at Medzamor, Armenian.
3. Research nuclear reactor was stopped in Georgia.
4. Nuclear equipment and different reactors placed in the European part of Russia.
5. WER-1000 nuclear reactor being built at Bushire, IRAN.
6. Technological equipment for nuclear material of Kazakhstan.
7. The purposed nuclear reactor with high-speed neutrons with vapor and pollution output, in Aktou, Kazakhstan.

M.K. Zaidi and I. Mustafaev (eds.), Radiation Safety Problems in the Caspian Region, 1-4.
© 2004 *Kluwer Academic Publishers. Printed in the Netherlands.*

Now, the project of the automated remote monitoring of environment background radiation in settlements along the boundary of Azerbaijan, and also along eastern suburbs of Azerbaijan regions inhabited by Armenia's army, was fully developed. The main purpose of the project is:
· Increase of a level of a radiation safety on territory of Azerbaijan,

· Controlling of a level of an environment and background radiation on boundary of the Azerbaijan with the purpose of well-timed warning and acceptance of indispensable measures at probable emergencies on Atomic Power Stations in a number adjacent from Azerbaijan countries, or other ecological catastrophes,

· Controlling a level of an environment background radiation along eastern suburbs of Azerbaijan regions occupied of Armenia's army and detection of the facts of wrongful disposals of atomic engineering waste generated by Armenia on territory of Azerbaijan.

As is known, in a number adjacent Azerbaijan countries the nuclear industry is advanced or being developed. It has resulted in origin of a threat of radiation hazard in case of ecological catastrophes: may be a wide scale leakage of radioactive wastes, explosions, or fires on nuclear generating plants, acts of sabotage, directional against Azerbaijan. In this case, at unfavorable meteorological conditions, a radioactive waste may be brought by a wind or a rain on territory of Azerbaijan.

Due to a huge atomic power station construction in Bushire (IRAN), it is necessary to estimate an initial background radiation along southern boundaries of Azerbaijan, that to determine probable changes in an ecology of southern boundaries during regular maintenance of atomic power station. Other one is an Armenian atomic power station, which is the padding center of a heightened radiation hazard in the Caucasian region. It is stipulated by several reasons:

· territory of Armenia is heightened seismic activity, therefore in case of earthquake there is a threat of destruction of generating sets of the Armenian atomic power station;

· because of absence of direct land transport communicational linkages, Armenia is not capable to remove waste products of atomic energetic to Russia, therefore it is rather probable, that these highly a radioactive waste dump on occupied territories of Azerbaijan;

· the atomic power station is the object, requiring of realization of regular, different repair - preventive actions, which highly qualified Russian and Armenian experts conducted one in the Soviet period. However after the collapse of the USSR scientific and technical links between scientists of republics were broke, and many experts leading in the field of atomic engineering have left the country. Therefore, as it is noted also by experts in Yerevan, Armenia, safety in operation of the Armenian atomic power station to be on a low level because of impossibility of realization of well-timed repair-preventive activities.

From this, realization of round-the-clock control of the environment conditions, first of all a background radiation check, along the boundary of Azerbaijan, and also along eastern remote areas is extremely a big problem.

PROCEDURE:

Measurement were done with the help of "EKOMON" fixed stations, consisting of appropriate sets of registration, the analysis and transfer of received data on an off-wire communication circuit and an autonomous supply set. The results of round-the-clock, gamma and neutron background measurements from the stations will be transmitted automatically to a dispatcher station in the central computer. On a dispatcher station the radiation mappings will be made automatically. Established on the stations telescopic sensors also will allow to determine a direction of a radiation and coordinates of radiation source. Stations will be located along boundary, and also in Kedabek, Akstafa, Terter, Agdam and Fizuli regions, and in Nakhichevan.

Once a month, the group of engineers goes around the stations by car and check up the instrumentation. One station will be established on the car, and during maintenance prevention activities at by-pass of stations will be made padding measurement of a background radiation along all boundaries of the above mentioned regions. Besides it, at realization of monitoring in the above indicated 5 regions through definite intervals of path at by-pass on the car samples of a ground and water will collect and operatively to be made an express radionuclide analysis on a carried gamma spectrometer. It will allow to define paths of migration of radionuclides in soil and water, and to define probable places of wrongful dumping of nuclear waste products.

DISCUSSIONS:

The mockup of the installation "ECOMON" is now developed and made, the application for the invention of the device of data transfer is conveyed and the positive decision is obtained, the software package for control of the installation and data processing in the computer is prepared.

Use of a method of the iterative designing, focused on use of RAD-resource and systems of automatic generation of initial texts on the basis of the created formal model it is made the software for remote control by the automated system of radiating monitoring.

On the basis of results of complex research of functional blocks of the developed system working breadboard model of the automated system of operative remote radiating monitoring is offered and made.

Use of mathematical model of formation of Vernandsk's anthropogenesis landscape it is developed a technique of realization of experimental measurements on the automated system of radiating monitoring. On the basis of the analysis of character of interaction of natural properties of an environment also it is artificial the created forms man-caused influences borders of a zone of sites of monitoring are determined.

Use of a new technique of realization of measurements executes experimental remote monitoring a radiating background of various territories of the Azerbaijan republic (see the map).

If necessary, it will be possible to equip the stations with padding sensors for measurement of some other ecological environmental factors: a level of an electromagnetic background, density dangerous and noxious gases in air basin, temperature, pressure, damps of atmosphere, a direction and wind speed. It will allow more precisely and to forecast development of an ecological situation not only in Azerbaijan but also in Caucasian region.

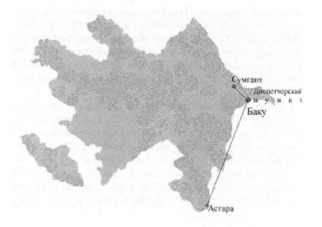

CONCLUSIONS:

The project, automated remote monitoring of the environment and background radiations in settlements along the boundary of Azerbaijan Republic was developed. Measurements were done with the help of "EKOMON" fixed stations, consisting of appropriate sets of registration, the analysis and transfer of received data on an off-wire communication circuit and an autonomous supply set. It will be possible to equip the stations with padding sensors for measurement of some other ecological environmental factors: a level of an electromagnetic background, density of dangerous and noxious gases in air basin, temperature, pressure, dampness of the atmosphere, direction and wind speed. It will allow us to forecast development of an ecological situation more precisely not only in Azerbaijan, but also in whole the Caucasian region.

REFERENCES:

1. Pashayev A.M, et.al. (2002). Remote monitoring of the environment on boundary of the Azerbaijan Republic. Int. Workshop "Effect of Ionizing Rad., Book of Abstracts 139.
2. Pashayev, A.M. et.al. (2003). NATO ARW, Baku. Abstracts Book. 28.
3. Makhonko, K.P. (1993). Radiation condition on the territory of Russia and bordering states in 1992. A yearbook of Obninsk – SPU "Typhoon".
4. Garibov, A.A. (2002). Influence nuclear technology systems of the nearest countries to radiation safety. Int. Workshop "Effect of Ionizing Radiation. Book of Abstracts, 11.
5. Maksimov, M.T., et.al. (1989). The radioactive pollutions and their measuring, Moscow, Energoatomizdat.

MONITORING OF THE RADIOECOLOGICAL SITUATION IN MARINE AND COASTAL ENVIRONMENT OF GEORGIA

AVTANDILASHVILI M.[A], BARATASHVILI D.[B], DUNKER R.[C], MAZMANIDI N.[C], PAGAVA S.[A,*], ROBAKIDZE Z.[A], RUSETSKI V.[A], TOGONIDZE G.[A]

[a]) Radiocarbon and Low-Level Counting Section of Nuclear Research Laboratory at Physics Faculty of I.Javakhishvili Tbilisi State University, Tbilisi, 0128, GEORGIA
[b]) Selection and Plant Protection Department of Batumi Botanical Gardens, Batumi, Georgia
[c]) Environmental Monitoring Laboratory, Department of Physics Idaho State University, Pocatello ID, U.S.A.
[d]) Sea Ecology and Fishery Institute, Batumi, Georgia

Corresponding author: spagava@access.sanet.ge,

ABSTRACT:

National Research and Educational Collaboration for Radioecology (NRECR), successfully acting in Georgia as independent expert institution since 1998, initiated studying of the radioecological condition in regions of Georgia and assessment of the risk for population from ionizing radiation. Analysis of results were obtained during 2001-2002 show soil polluted by anthropogenic radionuclides, committed annual individual effective dose due to external exposure for population come to 0.15-0.35 mSv; ^{137}Cs levels in agricultural products sometimes exceeds the recommended values.

Keywords: Ecology, radiation, contamination, survey, data collection and analysis.

INTRODUCTION:

Today the issue of the peaceful co-existence into the ecologically pure environment is placed in the center of the progressive world community's attention. In connection to this and taking into consideration the current reality of the Southern Caucasus and particularly Georgia, studying of the radioecological condition, on the one hand, and assessment of the potential risk for population's health, caused by the ionizing radiation, on the other hand, is extremely topical. At the same time it is to be noted that after the independence restoration the officials of Georgia – the Southern Caucasian Newly Independent State (NIS) – have not properly established the national infrastructure of the environmental radiation monitoring and are still unable to elaborate and carry out successfully other comprehensive environment-saving programs.

Investigation of the radioecological condition and potential risk assessment is important because of the grave inheritance – environment (soils, surface and ground waters etc.), polluted by the biologically active radionuclides with long half-life – received by Georgia and other newly post-communist states. This grave inheritance is caused by the Chernobyl Accident and the testing of

M.K. Zaidi and I. Mustafaev (eds.), Radiation Safety Problems in the Caspian Region, 5-12.
© 2004 *Kluwer Academic Publishers. Printed in the Netherlands.*

6

nuclear weapons in the atmosphere. Specifically, after the nuclear weapons testing in the atmosphere carried out in 20th century 60s in the environment of Georgia, as of the Northern hemisphere, the concentration of the radiocarbon (^{14}C) increased like tremendously [1, 2].

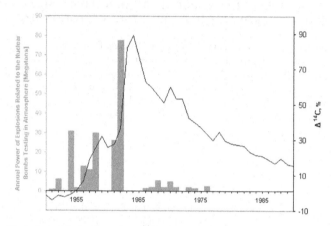

Fig. 1: Variation of the radiocarbon (^{14}C) concentration in Georgian collection wines (correspondingly, in atmospheric CO_2) in 1950-1990.

Radioactive fallout after the Chernobyl Accident caused significant increase in soil upper layer pollution at the territory of Georgia by various radionuclides with long half-life [3] (Fig. 2.).

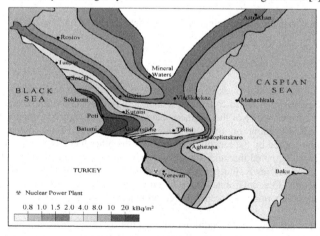

Fig. 2: Pollution of the Southern Caucasian territory by ^{137}Cs in 20th century 90s.

Moreover, it is to be stressed, that high concentration of radionuclides is observed at the territories of the former Soviet military sites and research institutions were involved in classified programs [4] The result of the joint survey carried out by Radiocarbon and Low-Level Counting Section (R&LLC) in cooperation with colleagues from Idaho State University, Pocatello, Idaho, USA) in 2003 confirms this statement (Fig. 3). These □-spectra were acquired on the territory of the Gonio fortress at the area opened by archaeologists in June 2003 (curve-1) and at the untouched area near the citrus plantation (curve-2), and on the territory of the former all-Soviet Institute for Tea and Subtropical Plants, at the building of Department for Radiation Genetics and Plants Physiology (curve-3).

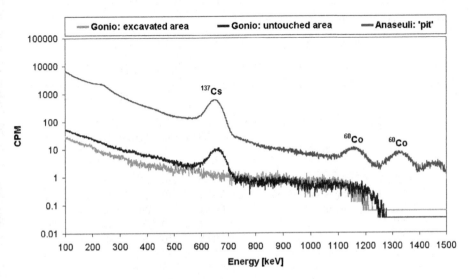

Fig. 3: Comparison of the □-spectra were acquired at various sites in Western Georgia

In 1998, R&LLC with the support from Georgian Branch of Soros Foundation (OSGF), the National Radioecological Research and Educational Collaboration formed the Collaboration with aim to study the radioecological conditions in region of Georgia and to assess the potential risk for population's health caused by the ionizing radiation.

RESULTS:

In 2000-2001, the Collaboration worked on investigation the radioecological condition in Chakvi-Sarpi section of the Black Sea coast. In particular, were studied the levels of soils and marine environment pollution by natural (^{40}K, U-Th decay chain members) and anthropogenic (^{137}Cs) radionuclides. Transport of radionuclides from environment to separate species of flora and fauna are consumed widely as foods in the region of investigation. The results of studies are presented in tables 1,2,3,4,5 [4,5].

Table 1: Natural and anthropogenic radionuclides in soil samples.

Sample ID	Sampling Place	Weighted Mean Activity [Bq/kg] dry mass							
		^{40}K	^{137}Cs	^{210}Pb	^{212}Pb	^{214}Pb	^{214}Bi	^{226}Ra	^{228}Ac
GAR-1	BBG Seaside	570±6	364±5	49 ± 9	13 ± 6	16 ± 5	17 ± 5	46 ± 9	<MDA
GAR-2	Gonio fortress	600±7	342±5	52 ± 8	14 ± 4	17 ± 6	< MDA	50 ± 6	<MDA
GAR-3	Batumi sea front	640±7	332±6	47 ± 9	19 ± 5	14 ± 6	< MDA	52 ± 8	<MDA
GAR-4	Chakvi	580±6	435±6	55 ± 8	21 ± 5	19 ± 4	23 ± 4	66 ± 6	33 ± 6
GAR-5	Leghva	590±6	388±5	48 ± 9	17 ± 5	17 ± 5	< MDA	63 ± 8	28 ± 7
GAR-6	Gonio	530±6	380±6	51 ± 9	20 ± 4	16 ± 5	< MDA	60 ± 8	30 ± 7
GAR-7	Kvariati	560±6	368±5	54 ± 8	18 ± 5	15 ± 5	< MDA	58 ± 6	<MDA
GAR-8	Sarpi	550±6	374±5	51 ± 8	16 ± 5	14 ± 5	< MDA	62 ± 6	<MDA

Table 2: Natural and anthropogenic radionuclides in vegetation samples.

	Sample ID	Sample Type	Weighted Mean Activity [Bq/kg] dry mass			
			^{7}Be	^{40}K	^{137}Cs	^{210}Pb
Batumi Botanical Gardens	GAR-20	Salad	270 ± 15	630 ± 25	15 ± 2	26 ± 5
	GAR-21	Fungus 'mantchkvala'	< MDA	810 ± 25	730 ± 20	21 ± 2
	GAR-22	Fungus 'pimpila'	< MDA	740 ± 20	850 ± 20	18 ± 7
	GAR-23	Grass	180 ± 8	670 ± 30	68 ± 2	86 ± 6
	GAR-24	Moss	< MDA	150 ± 20	380 ± 25	14 ± 7
	GAR-25	Gingko biloba leaves	73 ± 9	330 ± 15	7 ± 2	< MDA
	GAR-26	Citrus leaves	280 ± 20	1100 ± 30	10 ± 4	65 ± 8
	GAR-27	Cucumber leaves	340 ± 20	870 ± 20	11 ± 2	51 ± 5
	GAR-28	Tomato leaves	270 ± 20	790 ± 30	21 ± 3	47 ± 5
	GAR-29	Tobacco leaves	360 ± 20	660 ± 20	15 ± 2	43 ± 5
Gonio fortress	GAR-30	Salad	230 ± 15	780 ± 25	13 ± 1	30 ± 6
	GAR-31	Fungus 'mantchkvala'	< MDA	750 ± 20	850 ±25	< MDA
	GAR-32	Fungus 'pimpila'	< MDA	680 ± 20	780 ± 25	17 ± 6
	GAR-33	Grass	120 ± 5	770 ± 43	87 ± 3	115 ± 8
	GAR-34	Moss	< MDA	140 ± 20	340 ± 20	15 ± 5
	GAR-35	Citrus leaves	240 ± 15	1250 ± 30	6 ± 1	50 ± 6
	GAR-36	Cucumber leaves	260 ±15	950 ± 25	12 ± 2	45 ± 5
	GAR-37	Tomato leaves	240 ±20	860 ± 20	16 ± 2	57 ± 6
	GAR-38	Tobacco leaves	310 ±20	710 ±30	11 ± 1	55 ± 6
	GAR-39	Kiwi leaves	110 ± 15	840 ± 30	13 ±2	48 ± 5

Table 3: Natural and anthropogenic radionuclides in the Black Sea biota samples.

Sample ID	Sample Type	Weighted Mean Activity [Bq/kg] ^{40}K	^{137}Cs
GAR-60	Fish – Mugil cephalus L.	83 ± 4	0.7 ± 0.2
GAR-61	Fish – Odontogadus merlangus euxinus N.	70 ± 6	0.9 ± 0.3
GAR-62	Fish – Mullus barbatus ponticus E.	47 ± 6	< MDA
GAR-63	Spawn – Mullus barbatus ponticus E.	93 ± 23	< MDA
GAR-64	Fish – Platichthys flesus luscus P.	42 ± 6	< MDA
GAR-65	Fish – Cottus gobio Z.	63 ± 6	0.6 ± 0.4
GAR-66	Fish – Salmo trutta labrax P.	95 ± 7	< MDA
GAR-67	Shellfish – Rapana thomasiana G.	48 ± 5	< MDA
GAR-68	Shrimp – Palaemon adspersus R.	68 ± 5	< MDA
GAR-69	Algae (dry mass) – Macrophyta	185 ± 17	< MDA

Table 4: Natural and anthropogenic radionuclides in soils and tea plant at the Western Georgia.

Sampling place	Sample ID	Sample type	ACTIVITY [BQ/KG] DRY MASS ^{40}K	^{137}Cs	^{210}Pb	^{226}Ra
Anaseuli	TB-2229	Soil	205 ± 17	549 ± 11	154 ± 30	101 ± 20
	TB-2231	Tea Leaves	299 ± 23	23 ± 2	< MDA	< MDA
Laituri	TB-2247	Soil	255 ± 15	278 ± 6	116 ± 29	70 ± 16
	TB-2248	Tea Leaves	468 ± 37	20 ± 2	< MDA	< MDA
Tsetskhlauri	TB-2249	Soil	180 ± 13	201 ± 5	115 ± 29	50 ± 16
	TB-2250	Tea Leaves	307 ± 38	15 ± 2	< MDA	< MDA
Leghva	TB-2234	Soil	246 ± 17	334 ± 7	167 ± 31	126 ± 19
	TB-2235	Tea Leaves	414 ± 25	8 ± 2	< MDA	< MDA
Chakvi	TB-2232	Soil	201 ± 17	538 ± 11	187 ± 34	101 ± 16
	TB-2233	Tea Leaves	312 ± 23	12 ± 2	53 ± 21	< MDA

Table 5: Distribution of the chromosomal aberrations in tea plant from the W.Georgia.

Tea plantations	Frequency of chromosomal aberrations, % 1980	1990	2000
Anaseuli	1.9 ± 0.3	4.7 ± 0.5	3.8 ± 0.3
Laituri	1.4 ± 0.2	4.1 ± 0.4	3.0 ± 0.4
Tsetskhlauri	1.2 ± 0.2	4.9 ± 0.5	3.2 ± 0.2
Leghva	1.8 ± 0.3	5.0 ± 0.4	3.3 ± 0.3
Chakvi	1.7 ± 0.3	5.4 ± 0.5	3.6 ± 0.4

Note: Validation of the R&LLC results according to international standards are guaranteed by the inter-comparison program "Ringversuch zur Bestimmung der Spezifischen Aktivität

Künstlicher und Natürlicher Radionuklide" was carried out in 1999-2002 in collaboration with the Federal Agency for Protection against Radiation, Germany (BfS, 2002).

To perform these tasks were selected:
1) Permanent observation sites:
  ❑ Area in Seaside park of Batumi Botanical Gardens.
  ❑ Area in citrus plantation on the territory of Gonio fortress.
  ❑ Areas of the Black Sea aquatory at Gonio-Sarpi and Batumi port.
2) Test-objects:
  ❑ Soils from Chakvi-Sarpi section of the Black Sea coast and samples of vegetation consumed widely as foods.
  ❑ Surface water, sediment and separate species of the Black Sea biota.

CONCLUSION:

In the region of investigation the clear tendency of "purification" of the upper layer of soils, polluted by ^{137}Cs as a result of fallout from Chernobyl Accident, the soil activity amounted approximately 500 Bq/kg was not observed till 1990 [6]. Average and maximal committed individual annual effective dose for population caused by ^{137}Cs amount to 0.15 mSv and 0.35 mSv correspondingly [7]. Concentration of ^{137}Cs in separate samples of the agricultural products, were selected as test-objects, exceeds the value indicated by regulations for radiation safety. In particular, fungus (Fig. 4) consumption as food by population of the region is to be limited in certain way, moss usage as cattle food is to be discontinued. In separate cases, concentration of ^{210}Pb and ^{226}Ra is also significant.

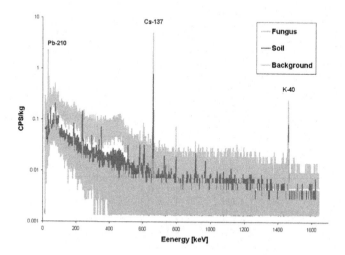

Fig. 4: Radionuclides selective "uptake" by fungus.

In agroindustrial tea plantations, the mutagenesis caused by ionizing radiation in tea plant increases as they absorbs various chemicals such as potassium (K) and caesium (Cs) from the soil. The observed concentration of radionuclides in leaves of citrus and tea plant is also to be drawn attention to. It is to be taken into consideration that leaves, polluted by radionuclides, can be a secondary source of contamination and in certain cases utilization is necessary. Samples of the Black Sea ichthyofauna were selected from pelagic (Mugil cephalus L., Odontogadus merlangus euxinus N.), benthopelagic (Mullus barbatus E.), benthic (Cottus gobio Z., Platichthys flesus luscus P.), migratory (Salmo trutta labrax P.) species. Other species of the Black Sea biota (Shellfish – Rapana thomasiana G. and shrimp – Palaemon adspersus R.) and algae (Macrophyta) were investigated too. Concentrations of ^{137}Cs are far below the regulation limits. Therefore, consumption of these marine products as food by population of the region is not restricted. It was also determined, that the distribution of ^{137}Cs in surface water near Chakvi-Sarpi section of the Black Sea coast is not uniform – the influence of river Chorokhi is observable: Content of ^{137}Cs in surface water near Chakvi-Gonio section decreases approximately by 10-15 % in comparison to Gonio-Sarpi section. Distribution of ^{137}Cs in bottom sediment layers in 1-10 cm interval is not uniform too. The influence of river Chorokhi is observable in this case as well. ^{137}Cs and other radionuclides penetrate to alluvium from drainage basin of river Chorokhi.

DISCUSSION:

Hence, aims and results of the presented investigation are in harmony not only with the aspiration of the progressive world community – to live and collaborate into ecologically pure environment – but are literally similar to recommendations, were formulated in Environmental Performance Review (EPR) of Georgia, issued by United Nations Economic Commission for Europe (UNECE) [8, 9].

Some recommendations from Georgian officials are posted below:

The Ministry of Environment and Natural Resources Protection should strengthen its Nuclear and Radiation Safety Service and identify sources of financing to:

(a) Further inventory and investigate all sites to provide detailed information on kinds of contamination and methods of rehabilitation;
(b) Speed up existing projects for the rehabilitation of contaminated sites.

The Ministry of Environment and Natural Resources Protection should:

(a) take appropriate measures to protect the population and to limit access to the Iagluji site.

The Ministry of Environment and Natural Resources Protection:

(a) should adopt the program on monitoring of industrial hot spots and high-polluting facilities should be included in this program as a matter of priority;

(b) should harmonize the local environmental norms and standards with international norms and standards, and should set up an appropriate system for environmental monitoring;

(c) should accelerate preparation of a Georgian national action plan for the Black Sea.

In accordance with western politologists' opinion, "Georgia, as other Southern Caucasian NISs, belongs to so named small countries. At the present stage of development of these countries are unable to eliminate the consequences of ecological disasters without active support and assistance from the developed countries [10].

REFERENCES:

1. Pagava S. (1989) Anthropogenic ^{14}C variations, Radiocarbon 31 (3) 771-776.
2. Radiation Doses, effects, risks. (1985). UNEP.
3. Kryshev I. (1992) Radioecological Consequences of the Chernobyl Accident. Moscow.
4. Pagava S. (2002) Study of Environmental Changes Using Isotope Techniques, IAEA. 480-481.
5. Pagava S. (2002) Radioecology Problems and Prospects. Environment and Radiation, No.5.
6. Ringversuch zur Bestimmung der Spezifischen Aktivität Künstlicher und Natürlicher Radionuklide in Filterschlamm. (2002) BfS, Berlin.
7. International basic safety standards for protection against ionizing radiation (1996). IAEA. ISSEP 07/01 Project Report. (2002) Tbilisi.
8. Environmental Performance Review of Georgia. (2003) UNECE.
9. Pagava S. (In press) Investigation of radiation condition in Chakvi-Sarpi coastal section and nearby aquatory of the Black Sea. Popularizing of radioekological studies. OSGF, Tbilisi.
10. Rondeli A. (1996). International Relations, Tbilisi.

ECOLOGIC SITUATION OF THE ARAL SEA REGION

E. Oteniyazov
Complex Institute of Natural Sciences, Karakalpak Branch,
Uzbekistan Academy of Sciences, Nukus, UZBEKISTAN.

Corresponding author: bakhyth@yandex.ru

ABSTRACT:

The paper deals with the analysis of the ecological situation of the Aral Sea region. The data on anthropogenic, which led to breakdown of ecosystem is sited in this paper. The rate of lowering of the Aral Sea level, the increase of water salinity and the influence of ecosystem breakdown on the health of the population are also mentioned in the paper.

Keywords: Ecologic disaster, Aral Sea basin, contamination, desertification, irrigation, mineralization and gamma radiation.

INTRODUCTION:

One of the most dangerous areas of ecological disaster in Central Asian region was created on drying up of the Aral Sea due to acute shortage of water resources. It also developed in an area of great pollution, desertification, aggravation of land resources and vanishing of biological resources, which caused threat for steady development of region. On the ecological and socio economic effect, the problem of the Aral Sea represents one of the largest disasters of 20th century [1]. Any ecological problem, including Aral tragedy, infringes interests of nation, therefore an attention of a majority of a public conversions to it [2].

Until the middle of 20th century the Aral Sea was the fourth in a world on the dimensions among self-contained pools. For many centuries, the two rivers – Amudarya and Syrdarya divided the water between an irrigation of arid eremic oases and Aral Sea [3]. Five independent states – Kazakhstan, Kyrgyzstan, Tajikistan, Uzbekistan, and Turkmenistan, and also Afganistan, are located on the territory of the Aral Sea basin. The basin located at the center of Central Asian deserts and is a giant vaporizer from which about 60 km^3 water per year entered in an atmosphere until recently, thus it was large climate-forming, temperature-controlling factor. Besides the sea was a huge receiver of salts, carried out in it by the rivers.

The Aral Sea and the Aral Sea region problems initially originated as a common ecological problem, at present have grown into the problem of man' s ecology, health and life in the conditions of anthropogenic desertification. The centuries-old stable ecosystem breakdown takes place. The high productive unique natural complexes become extinct, the priceless natural resources-genofund of endemic flora and fauna are irretrievably lost [4].

M.K. Zaidi and I. Mustafaev (eds.), Radiation Safety Problems in the Caspian Region, 13-16.
© 2004 *Kluwer Academic Publishers. Printed in the Netherlands.*

MAIN PART AND DISCUSSION:

During a conditional - natural state the water level in the Aral Sea (1910-1955) was on an absolute mark 53.2 m, a volume was 1060 km^3 at the surface area of 66 thousand km^2 and mineralization of 10-11 g/l. In 1987-88 the sea divided to two parts – big sea on the South and small sea on the North, at the water level 38.5 m (abs). In 1990, the water level in the Aral Sea has fallen on 16 m in last thirty years, and reached a critical mark of 36.6 m, the area of the sea reduced to 52%, a water volume decreased to 79%, and a salinity of sea water increased to 35-36%, which is almost corresponds to that of the global ocean. There is a major reduction in fish and transportation industries.

At the present time, the sea level has decreased to more than 20 m (abs), the area reduced 2-3 times, volume of the basin reduced almost 6 times, salinity in some places exceeds 100 g/l, and the whole of west basin of big sea has the salinity about 80 g/l. Water of Amudarya and Syrdarya entered in the sea in 1998 and 2003 only.

On the drained bottom of Aral Sea, sandy-salt deserts of more than 38 thousand km^2 were formed, called by people as "Aralkum". According to an estimate, toxiferous salt-sandy-storms, ascended from dried bottom, carry on Central Asia annually more than 70-80 million tons of toxiferous salts and dust [5,6].

Desertification of the Aral Sea region is accompanied by lost of land and ground resources, worsening of the quality of pastures and hayfields, decreasing of geno-pool of natural flora and fauna. More than 200 kinds of plants and animals have vanished and some of them become infrequent.

In conditions of a continued decrease of a level of the Aral Sea and development of processes of anthropogenic desertification in the region salification of soils happens faster and covers new light-salted or not salted areas, that has called falling fertility of dabbled grounds and reduction of biovariety.

In process of departure of the sea the area of brackish-sand desert is increased and there are intensive salt-accumulation take place. It has entailed gang of moisture-loving greens on saliniferous, and then on wilderness. In opinion of our scientists, the biological efficiency of the Aral Sea region as result of an anthropogenic desertification has decreased by 10 times: moving-pastoral holdings have disappeared, tugai and shrubby greens are perishing.

Disadvantage of water and drainage of the large areas of pools have reduced and crashed the fish industry. The republic extracting at the end of 1950 and beginning of 1960 up to 250 thousand centers of fish, in past year has extracted only 2 thousand centers. Nowadays in the Aral Sea in limits of Uzbekistan there are no fish [7,9].

Poor quality of drinking water, the contaminating of foodstuff by toxicant chemical materials (pesticides, herbicides, defoliants) results in different diseases, anemia, contagious, allergic etc of the population of region. The women of childbearing age and children are especially vulnerable.

In region of the Aral Sea one of the high levels of a case rate by a tuberculosis in terrain of Eurasia and one of the high levels of a case rate by an anemia all over the world [8].

About a quality of drinking water: The research conducted so far indicates the presence of ions of heavy metals and traces of elements, such as: Ca, Fe, F, Br, Cu, Zn, Mn, Ti, Ba, and Mo, are fixed in drinking waters of lower river Amudarya. On their data the contents of Fe on all districts is 3 times less than maximum permissible concentration (MPC), the concentration of fluorine exceeds 1.6 times in well and tap water in some districts, and in a number of districts it is 3.3 times less than MPC. Concentration of Br is 7 times less than MPC. Concentration of aluminum in well water of Muynak region is more than MPC by order, but in other tests the concentration of this trace element is 16.6 times less than MPC. Concentration of Ba in all tests exceeds MPC by 50 times, Zn is 70 times less than MPC, Co is 5 times of the MPC in tap water of two regions, and contents of Mo exceeds MPC in 4.4 times in tap and well water of two regions. It should be noted that all water test contents had small amount of iodine.

At present the area of drained bottom of the Aral Sea is about 38 thousand km^2, and it became one of the largest sandy-brackish area of Eurasia from which descends carrying-out the salty-dust to the environment. During 1960–1990, the carrying-out of the salty-dust was from 18 to 47 ton per hectare, and that from drained bottom propagates and settles as aerosols to adjacent territories on 150-200 thousand km^2. The vertical poles reach 200-400 km in length and 30-40 km in width, and repeat in large scales 10, and sometimes up to 15 times per year. Basic components of salty-dust are magnesium sulfate, calcium bicarbonate, sodium chloride etc., of which the last two salts are extremely toxic.

The unfavorable influence of different atmospheric contaminating, as a dust of a different parentage to an organism of man is well-known. For each state of basin of the Aral Sea the limits on a fence of water are fixed. For non-observance of fixed limits nobody carries of any responsibility. And though at apportioning of water the Aral Sea was chosen as separate water consumer, practically it never received full quota assigned. The Aral Sea basin area in many respects is defined by the consumer attitude to water, the discharge and use of river water from Amudarya and Syrdarya are based on selfish interests and the nobody is defending in the greater interest of the Aral Sea. At the existing conditions the loss of the unique sea, as a natural pool, will take place in the near future: it will divide in two - East and West. Formed on the dried bottom of the sea sandy-brackish desert "Aralkum" will soon reach the area more than 100 million hectares and more than 80 million tons of a salty dust annually will rise from its surface and carry far beyond the region of Aral Sea.

For softening ecological strength of the Aral basin the realization of complex measures on desertification processes control is envisioned: the Aral Sea level stabilization, creation of a system of mutually advantageous usage of transboundary waters, increase of efficiency of irrigative systems and dabbled fields, stopping of pollution of a water resources and preservation of quality of waters, creation and development of a system of a qualitative drinking water-supply and recognition of the relevant value of water, land and biological resources, as fundamentals of steady development of the region. The solution of these priority problems, according to our opinions, can promote softening of negative consequences of ecological crisis and supply long-time recovery of stabilization and way out to steady development of the Aral Sea region.

Conditions of a Radiation Safety of the region: The Republican Center of State Sanitary - epidemiological inspectors of Karakalpakstan implements the control behind a condition of a radiation safety on territory of the republic. The maintenance of a radiation safety from effect of ionizing radiations stipulated by radioactive matters environmental pollution, is reached by performance of the sanitarian legislation requirements. Radiation monitoring of a natural gamma radiation level is performed on the whole territory of Karakalpakstan, and also the radiation monitoring of imports of food and other materials is made.

The level of natural gamma radiation of the motorways with heavy traffic, residential areas, airport, and railway stations varies within $8 - 12$ µr/hour. In 1994 and 1999 the sanitary-epidemiological service was observed and performed radiation monitoring at the island Vozrozhdenie. The level of natural gamma radiation in the residential zone varied from 9 to 18 µr/hour. Only 5 units are using the sources of ionizing radiations in the Republic. These units have the sanitary passports on the right of storage, transportation and use of sources of ionizing radiations. The personal working with the radioactive substances and sources of ionizing radiations, drawn into centralized radiation monitoring, and among them the overflow of the specifications (SanPN – 0029-94) is not revealed [10].

CONCLUSIONS:

Elaboration of complex program of improvement of sanitary conditions and health status of the population in the ecological disaster zone seems to be an urgent issue. It should coordinate national economy, hydroeconomic, social and health care activities and prioritize health interests of the existing and future population.

REFERENCES:

1. Reimov R. (2001). Ecologic situation become worse. Vestnik of KK Branch of UzAS. 1-2.12
2. Bakhiev A, Konstantinova L.G. (2003). Newspaper articles. Vesti Karakalpakstana.
3. Abdirov Ch.A, et.al.(1996). Quality of waters of lower Amu Darya.Tashkent, FAN. 110
4. Reimov R. (1992). Ecological Problems of Aral: State of the Natural Environment and Genofund. Vestnik of KK Branch of UzAS. 4. 3-25
5. City of XXI Century. (1999). Ecological Anthology. Alma-Ata. 230
6. Konstantinova L.G. (1993). Functioning of Bacterial Assemblages of the Southern Aral Sea Region Pools in Conditions of Anthropogenic Effect. Doctoral Thesis, Tashkent. 49
7. Reimov. R. (1994). Mammals of the Southern Aral Sea Region, their Protection and Rational use in Conditions of Degenerative natural Environment. Ekaterinburg. C.48
8. Medical-ecological Situation in the Republic Karakalpakstan, Nukus. (1996). 18
9. Eschanov T.B, Azhibekov M.A. (2000). Physiological meaning and mechanisms of organism stressful reaction at adaptation. Vestnik of KK Branch of UzAS. 4.3-6
10. The concept of sustainable development of Rep.Uzbekistan. Tashkent (1998). 21.

DISTRIBUTION OF HEAVY METALS IN THE SOIL OF INDUSTRIAL ZONES OF APSHERON PENINSULA

L. A. ALIYEV

Institute of Radiation Matters, National Academy of Sciences
Ministry of Ecology and Natural Resources
159 Azadlig Avenue, AZ1106, Baku, Azerbaijan

Corresponding author: monitoring@mmd.baku.az

ABSTRACT:

A decline in industrial pollution by heavy metals at plants of chemical, petrochemical, machine building industries, thermoelectric power station, concrete plant, asphalt-concrete plant and "Electrocentrolit" plant in Baku and Sumgait was observed till 1994. The second notable decline of heavy metals contents was observed in 1999. For the last 3-4 years, the concentrations of heavy metals practically remained unchanged. A total decline of concentration in high levels of copper, zinc, nickel, and lead was noticed.

Keywords: Heavy metals, soil pollution, sources of soil pollution, radioactivity, heavy metals in soils of industrial areas.

INTRODUCTION:

In the result of agricultural human activity the pollution of the environment with various chemical hard, liquid, gaseous and organic wastes of industry, agricultural production, detergents, natural and artificial products of nuclear fission etc., the study of environment pollution with heavy metals is topical question itself. This study on the background of combined pressure by other factors on the environment seemed to be more interesting and urgent issue.
As it is known, metals with relative atomic mass more than 40 are heavy metals. Conception that availability of heavy metals in environment is an indispensable toxic factor is mistaken [1]. In microquantity heavy metals are principal environmental requisites, and some of them are elements of animate nature functioning. Thus, positive biological significance for elements to be included in this group-copper, zinc, cobalt, manganese, iron has been proved long ago. Some of them are important in agriculture in definite concentrations, i.e., toxicity of heavy metals is connected with certain concentrations exceeding Maximum Concentration Limit (MCL) [2,3,4].

Heavy metals are inhibitors of ferments in human organism [5]. Lead compounds impair metabolism, and the most dangerous consequences of non-organic lead compound influence is its ability to replace calcium in bones. Being constant intoxication source within long-term period, effect of lead on small children is especially dangerous it causes mental deficiency and chronic cerebropathy. Cadmium penetrating into human organism and continuously accumulating in organs, leads to chronic intoxication [6].

M.K. Zaidi and I. Mustafaev (eds.), Radiation Safety Problems in the Caspian Region, 17-22.
© 2004 *Kluwer Academic Publishers. Printed in the Netherlands.*

Specific disease (ita-ita disease), revealed in Japan is called cadmic disease. In experience of this country cadmic pollution source was mining complex. This disease becomes apparent in dysfunctions of the organism in the result of osteomalacia [1].

Supply sources of toxins into lithosphere, especially into its upper soil conditionally may be divided into 2 groups: natural and man-caused. Among natural sources the following contribute mostly: rock weathering, erosion, volcanic activity, dust, fogs, volcanic gas. On such territory like Azerbaijan characterized by high volcanic activity, where number of active volcano run up to more than 200, nevertheless the most dangerous are polluting components of man-caused nature [6].

In the process of industrial activity man promotes the intensive dispersion of heavy metals in environment. Metals matriculate into atmosphere in the composition of gaseous secretions and smokes, and as man-caused dust; they enter with sewage into ponds, from water and atmosphere pass into the soil where their migration processes are slowed down because of strong conservatism of soil environment [7]. For this reason constant supply even in small doses may cause significant accumulation of metals in soil. Metals entering and precipitated in soil depends on the human activity in this region. Thus, source of lead supply into the environment are treated gas of internal combustion engine, of cadmium are phosphates used in agriculture, containing this element as admixture and widely-used compounds of this metal in and varnish-and-paint industry.

Supply sources of chrome into environment can be sewage of horology, tanning and heavy industries. Zinc passes into environment at systematic use of it as organic fertilizer of city sewage sediments, and at rubber wastes incineration. Atmosphere notably influence on the content of various elements in soils.

Heavy metals are mainly brought into atmosphere in the solution of aerosols that are of great importance in chemical pollutions of the air. It may contain lead, cadmium, arsenic, mercury, chrome, nickel, zinc and other elements [8]. Radionuclides, forming at nuclear fuel fission and outcropping on the earth surface in the petroleum production they also gets into atmosphere in disappearing small amounts and are of great ecological interest not for its toxicity, related with chemical properties of the elements, but for its radioactivity. Entering from atmosphere into soil and water within the long-term period, practically without changing the concentration of the elements in these surroundings, they may change background level of radioactivity and emanation nature [9].

As per research data, for the past years noticeable increase of natural radionuclides in biosphere is observed, the supply source of which are usually some types of mineral fertilizer, wastes of NTT and thermoelectric power stations, lay waters of oil and gas fields, mineral waters etc. Hence the conclusion is drawn that at total pollution of environment, including number of soil to be polluted by production wastes, radioactive pollution occurs. By this reason the revealing of anthropogenic influence on the soils requires complex approach to the problem:

1) Inventory of anthropogenic factors, directly or indirectly affecting the environment.

2) Assessment of summary power of influence of anthropogenic factors on naturally-climatic conditions of the region.

3) Assessment of quantitative heterogeneity of anthropogenic loading within the borders of this area.

To clarify this and other matters on urbanization influence on the environment quality we carried out long-term investigation of pollution with heavy metals of soils around the industrial enterprises and its correlation with γ-survey of the same territories of Apsheron peninsula cities of Azerbaijan.

Besides the natural habitat boundaries of man-caused pollutions in the industrial enterprises zones and the direction of influencing factors distribution were established. Also we continue the research of aerosols content (solid particles, weighing in air) in size from 0.1 - 10 micro-m and more (carriers of the basic mass of heavy metals in atmosphere and precipitating after 5 days or in 3-4 weeks) and germs (trophic bacterium group) in soils.

EXPERIMENTAL:

We were carrying out works on defining the availability of heavy metals in soil on the areas of petrochemical, chemical, machine building, metallurgical plants and thermal power stations in Baku and Sumgait. Content of heavy metals in soil were determined on atomic absorptive spectrophotometer [10]. Before proper analysis on atomic absorptive spectrophotometer metals were abstracted by nitric, sulphuric and chloric acids or their mixture, without breaking the silicate foundation of soil. Optimum alternative is the use of hot 1n. HNO_3. Preparation of standard solutions and reagents:

 ❑ Single-norm nitric acid was prepared by dilution of 1 dose (scopes) of concentrated HNO_3 in 15 dose (volumes) of twice-distilled water.

Basic standard solutions of metals with concentration of 1mg/ml were prepared as follows:

Zinc. 1gr of zinc is dissolved in 30 ml of hydrochloric acid with 1:1 concentration.
Copper. 3.798gr of $Cu(NO_3)_2$ $3H_2O$ was diluted in 250 ml of twice-distilled water
Cadmium. 11423gr of CdO was diluted in 100 ml of twice-distilled water
Lead. 1.598gr of $Pb(NO_3)_2$ was diluted in 100 ml of twice-distilled water.
Nickel. 4.953gr of $Ni(NO_3)_2$ $6H_2O$ was diluted in 100 ml twice-distilled water.
Chrome. 7.693gr of $Cr(NO_3)_2$ $9H_2O$ was diluted in 250 ml of twice-distilled water.

Received solution of metals was completed by twice-distilled water to 1 l Air-dry soil was reduced to powder state. 5 gr dose of milled soil were placed in conic glass retorts of 75-100ml capacity and stained with 30 ml of 1n. HNO_3. Retorts were placed on the heat block and the contents were heated to boil till almost dry evaporation of solution was obtained in an hour. 40 ml of 1n. HNO_3 was added to the residue and heated thoroughly for 5-10 min, and then it was filtered through the paper filter with blue band. Filter was accurately washed by 100 ml of 1n.

HNO_3 to remove metal remnants. Sediment on filter was washed by 1n. HNO_3, bringing filtrate volume in volumetric flask to 50 ml.

Then optic density (absorbance) of obtained solutions were defined on AA-spectrophotometer and metal concentration was determined by gage curve (standard curve). In soil of industrial areas and around the plants the availability of following metals was defined: nickel (Ni), cobalt (Co), lead (Pb), chrome (Cr), zinc (Zn), copper (Cu), tin (Sn), cadmium (Cd), vanadium (V), mercury (Hg).

DISCUSSION:

The average concentration of heavy metals in soil of industrial zones in Baku and Sumgait have been determined. By the contents, the metals of these soils are strictly distinguished into 2 groups:

1) heavy metals with peak concentration are about 40-65 mg/kg
2) heavy metals with peak concentration are hardly 10- 25 mg/kg.

In soils of industrial areas of Baku, group 1 includes copper, zinc and nickel; group 2 – cobalt, lead, vanadium, mercury, cadmium and tin. In Sumgait similar groups are formed accordingly, group 1 has zinc, copper and mercury and group 2 lead, cobalt, chrome, vanadium and cadmium.

Alternation trend of heavy metals in soils of Baku and Sumgait industrial areas are of similar nature: abrupt decline since 1991 up to 1993-1994, then second decay 1998-1999 and stabilization at definite level. It can be explained by similar level of production facilities activation in both cities. Changing tendency of contents between two abovementioned groups are of different character. Thus, metal content has declined within 1991-2002 abruptly and have stabilized on the level 20-22 mg/kg (exception among group 2, metal leads its content in soils of industrial zones, since 1998 has been fluently increasing, what can be explained by increasing number of auto transport working on gasolene. Such state may be clarified by the fact that factor of group 1 metals in soil in peak years was mostly connected with industrial pollutions, and metals of group 2 by Clarke of these area.

The dependence of the content from the distance has different nature. As it is seen maximal value of group 1 metals at various distances varies according to source of pollution. Since Cu concentration at the industrial area soils is observed at 1, 1-5 km from the source. But they are maintained within 50 km. Maximal Ni concentration is observed at 1, 1-20 km, and its insignificant decline at 20-50 km distance. Maximal magnitude of zinc concentration was different at the 0-1 km distance from industrial area, at the distance of 21-50 km it reducing to 12-15 mg/kg. Through the heavy metals, we investigated only lead, the contents were not distinctly depending on the distance, what is explained by its far location from transport not from industrial plants.

Soil samples were taken from various directions from pollution source at 20 km distance. Long-term information of similar samples for heavy metals analysis reveals fuzzy regularity. It was clarified that large concentrations accumulate in west direction. This state is explained by strong sea breezes.

Distribution of heavy metals was found in soil around plants of machine building, metallurgy, oil and chemical fields. Comparative analysis of received data shows that there is maximal content of zinc, copper and nickel in these soils. Depending on plant specializations the observed values differ. Thus, in 1991, areas were more polluted with heavy metals, maximal zinc concentration was observed in soils of machine building and oil-refining plants, Sumgait Thermal Power Station, abundant low content of zinc was noticed in soil around Karadag Concrete Works, Baku thermal power station, "Superphosphat" and "Synthetic kauchuk" Sumgait plants (60 mg/kg).

Relatively high values were revealed in soils of machine building and oil-refining plants, low values in soils at plants of chemical industrial (Sumgait plants - "Synthetic caoutchouk", "Additives" (Prisadki), "Synthetic detergents" and Thermal Power Stations). Comparative analysis of nickel contents in soil of industrial plants gave the following illustration: the highest values revealed in soils of Tube–rolling mill, "Synthetic detergents" plants and "Electrocetrolit".

Relatively low concentrations were discovered in soils of machine building, Sumgait chemical, petrol-refining plants, Karadag Concrete Works, Baku Thermal Power Station. On the areas of all industrial plants γ-background was at acceptable level.

CONCLUSIONS:

1. Soils of industrial zones in Baku and Sumgait cities are moderately polluted with heavy metals.

2. Dependence of heavy metals in soils at the distance of pollution source for various metals has different nature.

3. Dependence of content in soils in various directions at the same distance for different metals has different nature.

4. Absence of correlation between γ-background and heavy metals in soils of investigated zones led to the following conclusions: radioactive isotopes among metal hard wastes of these plants are not available.

REFERENCES:

1. Alexeev J.V. (1987). Heavy Metals in Soils and Plants, Leningrad, Agropromizdat.
2. Garmash G. A. (1985). The Accumulation of Heavy Metals in Soils and Plants around Metallurgy Plants, Auto Synopsis of PhD thesis. Biol. Sciences.
3. Brookes P. C., Mc Grath S. P. (1984). Effects of Metal Toxicity on the Size of the Soil Microbial Biomas, J. of Soil Science 35(2). 341-346.
4. Japanese Standards Association. (1998). Environmental Technology, Tokyo.
5. Abdullaev M. A., Aliev J. A. (1998). The Migration of Natural and Synthetic Radionuclids in Soil-plant Systems, Baku. Elm

6. Nikitin D. P., Novikov I.V. (1980), Environment and Human, Moscow, "Higher School".
7. Prokhorov V. M. (1974). The Migration of Radioactivity Pollution in Soils: Auto Synopsis PhD Thesis. Chem. Sciences, Leninqrad.
8. Novikova M. I., (1989), Monitoring of Environment, Moscow, 10-17.
9. Drichko V. F. (1983) Behavior in Environment of Natural Heavy Radionuclids. J Results of Science and Technique, Series. Radiation Biology. 66-98.
10. Khavezor I, Calev D.(1983) Atom Absorbsion Analysis. Leninqrad, Chemia.
11. Japparova J.M., Sadicova G.D. (2002). Distribution of Cadmium and Lead agile forms in soil of Alma-ate City. Hydrometeorology and Ecology 2.

STRUCTURAL PECULIARITIES OF BOROSILICATES AND ITS CORRELATION WITH RADIATION-CATALYTIC ACTIVITY

A.M. GASANOV, E.A. SAMEDOV, S.Z. MELIKOVA
Azerbaijan National Aerospace Agency
Institute of Radiation Problems, National Academy of Sciences
159 Azadlig Avenue, AZ1106, Baku, AZERBAIJAN.

Corresponding Author: sevincmelikova@yahoo.com

ABSTRACT:

The structural features of borosilicate are investigated with various percentage of B_2O_3 by Infrared (IR) spectroscopy and the Differential-Thermal Analysis (DTA) methods. Comparing the found lows with the data on a power output of non-equilibrium carriers of a charge on 100 eV the absorbed energy. The essential role of the three dimensional ions of pine bore in formation of the active centers participating in process radiolysis is made.

Keywords: Radiation center; borosilicate; three-dimensional and four-dimensional ions.

INTRODUCTION:

The radiation-heterogeneous processes on oxide surface and silicate catalysts are constantly investigated and are at the center of attention. Plenty of work is devoted to radiation-heterogeneous transformation of molecules of water and the hydrocarbons adsorbed on a borosilicate surface [1-3]. As it is known, the output of products of decomposition is closely connected to accumulation of the radiation paramagnetic center in SiO_2/B_2O_3, thus the borosilicate contains and the structure can vender the essential contribution to efficiency of the process. So an attempt of revealing interrelation of borosilicate structural features with efficiency of accumulation of the radiation paramagnetic center is undertaken.

METHOD:

Research carried out with powder silicate, which was synthesized by method of sedimentation. The quantity B_2O_3 in SiO_2 made 0.9, 1.5, 3, 10, and 50%. The structure borosilicate with different maintenance B_2O_3 was investigated by methods of differential-thermal analysis (DTA) and infrared (IR) spectroscopy. DTA curves were recorded on derivatograf by firm "MOM" in temperature diapason 20-1000°C. For revealing the structural features of borosilicate the quantitative analysis method with the help of IR spectroscopy as been used in the field of 4600-6500 cm^{-1}. Spectra have been received by pressing together with KBr in identical conditions. Pressing was made at a pressure of 10-65 kg/mm^2.

M.K. Zaidi and I. Mustafaev (eds.), Radiation Safety Problems in the Caspian Region, 23-27.
© 2004 *Kluwer Academic Publishers. Printed in the Netherlands.*

RESULTS AND DISCUSSIONS:

On DTA curves endoeffects fusion (T~370°C) are observed (Figure 1): with increase of percentage B_2O_3 in silicates broad and displacement of a maximum endoeffect aside high temperature takes place. At some concentration B_2O_3 (~1.5 %) three clear endoeffect the fusion related to various structures of borosilicate are observed. In high-temperature area, at 830°C exoeffects of crystallization take place. With increase of percentage B_2O_3 the maximum exoeffect is displaced to the low temperatures sides. Curves DTA SiO_2 (a) and borosilicates with various percentages; b - 0.9 %; c - 1.5%; d -10%; e - 50 %.

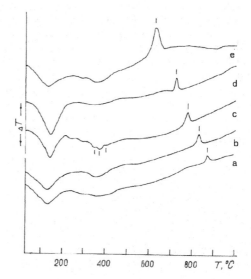

Figure 1.

While investigating of crystallized samples, exoeffects disappear. Observed lows are characteristic for silicate systems investigated earlier and are specified complex structural dependence borosilicate from their structure. On Figure 2, IR spectra SiO_2 and samples borosilicate with different percentage B_2O_3 are shown. As it is seen from figure initial SiO_2 the band 1095 cm^{-1}, corresponded to fluctuation of Si-O is shown in a spectrum. With increase in maintenance B_2O_3 the band is displaced in low frequency area. Displacement of this band in smaller frequencies is caused by replacement of atoms of silicon by atoms of a pine bore in silicon-oxygen tetrahedrons. Thus it is necessary to take into account, that frequency of valence fluctuation B-O is lower than frequency of fluctuations of communication Si-O, therefore with increase in number of Si-O-B, fluctuations Si-O are displaced in smaller frequencies. With change of structure borosilicate it is necessary to expect change of coordination of ions of a pine bore.

It is known; that ions of a pine bore in silicate can have three or four-dimensions [4,5]. The increase in dimensional number of ions of a pine bore entails the increase of B-O interatom distance that affects reduction of frequency of valence fluctuations. Therefore in IR spectra of silicates, it is possible to estimate presence of some kind of structure in borosilicate by intensity bands of absorption of corresponding BO_3 and BO_4 – grouping. According to the previous works [6] the absorption of BO_3 grouping in silicate is revealed at 1300 cm^{-1}, and absorption BO_4 grouping at frequency of 980 cm^{-1}. For a quantitative estimation of participation of these groupings in borosilicate structure we calculate relations of optical density of absorption bands of corresponding structures to optical density of absorption Si-O strip (Table 2):

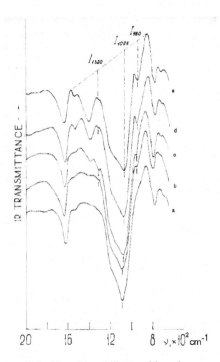

Figure 2. IR absorption spectra SiO_2 (a) and borosilicate with various percentage; b - 0.9%; c - 1.5%; d -10%; e - 50%.

It is shown from the table, that at maintenance B_2O_3 in structure of silicate of equal 1.5% the maximal maintenance of the three-dimensional ions of a pine bore is observed. Kinetic laws of accumulation of the radiation paramagnetic centers formed under action of □-quanta in borosilicate with different percentage B_2O_3 have been earlier investigated.

Table 2. Relation of optical density of absorption bands three-dimensional and four-dimensional pine bore to optical density of a absorption Si-O strip

№	%B_2O_3	D_{1320}/D_{1095}	D_{980}/D_{1095}
1	0.9	0.24	0.30
2	1.5	0.28	0.30
3	3	0.21	0.33
4	10	0.20	0.33
5	50	0.21	0.32

On Figure 3, the dependence of an output of non-equilibrium carriers of a charge per 100eV is submitted on borosilicate is contained. As it can be seen in the figure the maximal output of non-equilibrium carriers of charge is observed at 1.5% B_2O_3 maintenance in SiO_2.

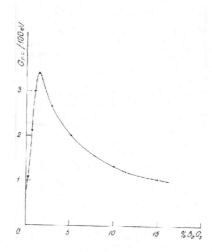

Figure 3. Dependence of a power output of non-equilibrium carriers of a charge on 100 eV.

CONCLUSION:

The essential role of the three dimensional ions of a pine bore in radiation-catalytic processes are a part of BO_3 of a triangular group with the atom of a pine bore located a little bit above a triangular plane. In this case, atoms of a pine bore in contrast to silicon atoms are charged negatively and when close to the surface acts as centers of local destabilization that participate in the process of radiolysis.

REFERENCES:

1.Samedov E.A.,Melikova S.Z. (2003). Proc.Conf.Nucl.&Rad.Physics,Almaty, KA.
2.Gadjiyeva N.N.,Samedov E.A. (1995). J. Appl. Spectroscopy. 62 (6). 44.
3.Garibov A.A., Samedov E.A. (2000). High Energy Chemistry, 34 (6). 421.
4. Broadhead P., Newman G.A. (1971). J. Mol. Structure 10. 157.
5. Samedov E.A.(2004). J. Appl. Spectroscopy 71 (1). 119.
6. Efimov A.M., Mikhailov V.A., Arkatova T.G. (1999). Phys. Chem. Glasses 5 (6). 692.

CHARACTERISTICS OF BROKEN GROUNDS AT THE FORMER AZGIR NUCLEAR TEST SITE

E. Z. AKHMETOV, ZH. I. ADYMOV, V. DZHEZAIROV-KAKHRAMANOV AND A. S. YERMATOV
Institute of Nuclear Physics, National Nuclear Center
Alamaty, 480082, KAZAKISTAN

Corresponding author: office@tandem-translations.com

ABSTRACT:

The phase and elementary composition of soil on the day surface of the technological sites of the former Azgir nuclear test site were studied and data collected on radionuclide contamination of soil in a number of sites and distribution of caesium-137 on grain-size fractions of soil has been presented.

Keywords: Azgir Nuclear Test Site, radionuclide, contamination and granulometrical fractions

INTRODUCTION:

Nuclear explosion experiments, carried out at the salt dome structure "The Azgir", introduced some changes in to the natural landscape of that locality and natural relief of the territories of locations. Explosion at Western Azgir exercised a strong seismic action onto the settlement "Azgir" in the form of destruction of dwelling and buildings, so further explosion experiments were transferred to Eastern Dome, at the distance of 20-25 km from the settlement "Azgir". At the location, funnel-shaped fall-through was formed as a result of non-optimal breaking ground for wells and because of accurate calculations, with an artificial water reservoir at the bottom, where the water is not useful for drinking and technical purposes as it contains minerals. During the carrying out of drilling after explosions and geophysical investigation of space, a stable contamination of soil ground, localized, mainly, by limits of technological locations territories around of benchmarks of combat wells. Also, at each location there we observed some lowering of the relief of the territory having the diameter of up to 200 m around the well. By measures on recultivation, the radiation situation has improved at technological locations, which, however, differed at some places, although in the less degree, in comparison with the post-explosion period, from the natural radiation background in the region, i.e. some increased values of both the power of exposure doses and contents of radionuclides in soil, caused by global fall-out and by output of artificial radionuclides from trench burials of radioactive waste. During recultivation, some radioactive spots were not liquidated, and now there exposure has taken place because of water and wind soil erosion and emission of radionuclides from those soils [1].

Remote after effects of underground nuclear explosions can be considered a continuing process of formation of ground falls-through at some locations of large-bottle-shaped form, of depth up to 3 m and diameter of up to 2.5 m. The manifestation of falls-through, was a consequence of

M.K. Zaidi and I. Mustafaev (eds.), Radiation Safety Problems in the Caspian Region, 29-32.
© 2004 *Kluwer Academic Publishers. Printed in the Netherlands.*

cast-formation in sediment over-salted rocks as well as of breakdown of environmental continuity as a result of nuclear explosion in the past [2].

METHODS:

The contamination of surface soil layer depends on the composition of soil ground, physio-mechanical and geological properties, structure and size of soil-formatting particles and conglomerates composed of those particles. The macro-composition of soil ground, phaseous components, was determined by the method of roentgen-diffractometry with use of usual experimental roentgen installations of the type "DRON-2". Sample of phaseon soil composition were taken from different locations, prepared using standard techniques, were analyzed. The roentgen-diffractometrical analysis was carried out with the use of the $\square$-filter. Conditions of survey by diffractogram were: $V = 35$ KV; the current, $I = 0$ mA; the scale – 200 pulses; the survey - $\square$ - 2 $\square$; the velocity of the detector movement – 2 degree/min. An interpretation of the diffractograms was carried out with use of "ASTM Powder diffraktum. They were pure without impurities of minerals. Possible mixtures, an identification of which could not be unambiguous because of small contents and a presence of only 1 - 2 – diffractional reflexes or bad crystalline, are presented on the diffractogram. A relative estimation of contents of roentgen – amorphous dispersion component (for small-iron probe) was carried out on the level of diffusional scattering in the region of small angles. An intensity of the background scattering at large angles of diffraction gave a relative estimation of total ferricity of the probes.

RESULTS AND DISCUSSION:

The analysis of experimental results shows, that composition of the probes "AZ$_1$" and "AZ$_2$" are identical there, where there are predominated gypsum, quartz and calcium. In both probes there is presented potassium feldspar. Besides, in the probe "AZ$_2$" there is a small amount of chlorite. In the probe "AZ$_3$", besides main components (quartz, gypsum, calcium, potassium feldspar and feldspar), gematite is present – the compound of iron with oxygen (Fe$_2$O$_3$). In the probe "AZy" a large quantity of quartz, potassium feldspar, feldspar, gypsum, calcium and small quantity of silicone, known as kaolinite Al$_2$[Si$_2$O$_5$]OH$_4$ and chlorite with chemical formula was present. (Mg$_2$Fe^{11}/Al, Fe^{111})$_{12}$$\square$ [(Si, Al)$_8$O$_{20}$]$\square$(OH)$_{16}$. In the probe "AZ$_5$", large quantities, quartz, potassium feldspar as well as feldspar and traces of mica were present. In probes "AZ$_6$"and "AZ$_7$", in different ratios, quartz, potassium feldspar, feldspar, calcium, mica and chlorite are present. In both probes the quantities of quartz were maximum, excepting "AZ$_1$" and "AZ$_2$", where there were dominated by gypsum. The composition of soil ground of technological location was determined with the help of roentgen-fluorescent analysis (RFA), which was carried out by the following scheme:

- a preparation of samples for the analysis;
- an excitation of a spectrum;
- a separation of analytical lines;
- a detection of lines intensities;
- an interpretation of measurement results.

The nuclide composition of samples was determined on roentgen radiation at the semi-conductor detector "GEM-2018" (ORTEC) with the detection efficiency of 20%. The soil samples were irradiated by radioactive source "Cadmium-109" and the detector was reliably shielded from the source and detected only roentgen radiation of soil samples. There determined the relative contents, in soil samples, of the following elements: K, Ca, Ti, Mn, Fe, Cu, Zn, As, Pb, Rb, Sr, Y, Zr, Nb, Mo, Cs, Ba, La and Ce. In the Table 1, results of composition of chemical elements from the locations "A1", "A5" and "A10", in which there are indicated limits of measured contents of chemical elements in samples from different plots of locations territories. Different values of elemental contents in soil at both one location and other ones show an essential inhomogeneity of elemental composition of present soil from technological locations. Such inhomogeneity may be caused by circumstances, that after selection and disposal of radioactive and chemical contaminated surface soil in underground space of "A10", places of excavation were filled up by clean soil brought from different places [3,4].

Table 1. Elemental composition of soil probes taken from locations of the Azgir Test Site

Loca- tion	Elements				
	K	Ca	Ti	Mn	Fe
	%				
A1	1.15-1.73	2.63-11.69	0.22-0.34	0.03-0.04	1.49-2.51
A5	1.51-1.77	3.82-5.79	0.30-0.35	0.03	2.54-2.56
A10	1.49-1.69	1.60-2.01	0.35-0.39	0.05-0.06	3.12-3.34
	Cu	Zn	As	Pb	Rb	Sr	Y
	g/t						
A1	8-29	29-50	8-37	3-15	33 - 60	215-517	5.6-13.5
A5	32-56	51-54	35-36	14	51-53	220-249	11.1-13.1
A10	28-29	69-83	35-36	14-15	67-69	206-259	16.7-17.4
	Zr	Nb	Mo	Cs	Ba	La	Ce
	g/t						
A1	104-187	4.8-8.3	1.5-3.2	10-45	458-1302	11-48	20-49
A5	176-198	7.6-8.1	1.6-1.7	40-42	296-300	45	46-47
A10	182-196	10.1-10.9	1.6-1.7	42-43	337-370	45-46	46-53

Granulometrical composition of soil from locations of the Azgir Test Site were used to study a soil particles blowing-out and a distribution of caesium-137 radionuclides. Our earlier works on granulometry of soil showed [5,6], that results of the distribution of soil particles on geometrical size depend, essentially, on the way of determination of particles quantity on fractions: dry or wet sieving. During the dry sieving, complete separation of small fractions has not take place; small dust-like particles (< 0.063 mm) were partly sticked together so there a distortion of results of granulometrical composition was observed as caesium-137 concentration on fraction. The use of the wet sieving allowed effectively to separate fractions and to measure more correctly the granulometrical composition and the distribution of caesium-137. In the Table 2, granulomenrical composition of soil 1 cm thickness and at a depth of 1 to 4 cm is presented. For every fixed soil fraction, there were limits of percent contents of soil particles, and probes were taken at different spots of the location and at different depths. The available spread in relative contents of soil particles in fractions are attributed by variety of phaseous composition and by that fact, that soil underwent significant technogeous changes of its macrostructure. The use of the dry method of sieving of soil particles had determined, that they were mainly concentrated in the fractions: 0.7 mm and 0.1 mm; and the use of water jet obtained more precise results – particles were mainly in fractions: 0.1 mm and < 0.063 mm. Both methods fixed together small quantity of particles in the fraction "< 0.063 mm".

Table 2. Granulometrical composition of soil layer at the location "A2", in percents

Way of soil processing	SIZE OF FRACTION, MM						
	1.25	0.7	0.4	0.1	0.063	<0.063	clay
Dry	0.4-4.0	12.3-34.1	21.6-25.4	28.3-48.4	3.7-7.5	4.6-9.9	-
Wet	0.1-1.2	0.2-0.42	0.2-1.0	22.0-40.5	6.7-10.2	44.6-67.6	0.6-3.9

In Table 3 there are given relative contents of caesium-137 in separated soil fractions, measured during the dry and wet ways of soil processing. By the dry sieving of soil particles that caesium – 137 was in the fractions: 0.7 mm, 0.4 mm and 0.1 mm; by the wet sieving there was determined that caesium-137 was concentrated in the fractions: < 0.063 (mainly) and 0.1 mm. Results of both ways of soil particles sieving gave small quantities of relative contents of caesium-137 in the fraction "0.063".

Table 3. Distribution of caesium-137 on soil fractions at the location "A2" in percent

Way of soil processing	SIZE OF FRACTION, MM						
	1.25	0.7	0.4	0.1	0.063	<0.063	clay
dry	0.5-3.5	13.2-31.	24.3-27.8	28.8-45.2	4.0-7.0	5.9-12.8	-
wet	0.0-0.6	0.0-3.1	0.0-4.0	3.6-21.9	1.2-5.8	58.9-92.2	0.7-7.4

CONCLUSIONS:

The soil cover at Azgir Test Site after carrying out of nuclear explosions and during the post-explosion period was subjected to an intensive negative technogeneous influence and experienced chemical and radioactive contamination of surface soil layer. Soil particles are mainly concentrated in granulometrical fractions [< 0.063 mm – (44.6-67.6)%, 0.1 mm – (22.0-40.5)%], and the radionuclide, caesium-137, is distributed in the fraction "<0.063" [(58.9-92.2)%], i.e. in small dust-like particles. It means, that caesium-137 is firmly connected with soil particles [5] and can be freely transferred with small dust-like fractions by winds; so, in the future, it is necessary to find some ways for biding of this fraction, and to avoid dust-transfer.

REFERENCES:

1. Yu.V. Dubasov. (1994) Analysis and stage by-stage of rad.situation.GALIT.NPORT. S.P.
2. V.B. Adamsky. (1998). Peaceful Nuclear explosions-"Big Azgir".10-98, INP NNC RK.
3. A.S. Krivokhatsky. (1993). Radiation manifestations. ZNIIatominform. M. № 3
4. A.S. Krivokhatsky. (1992). Main characteristics of rad.situation. I-223. ZNIIatominform.
5. E.Z.Akhmetov. (2000).Distributions of cs-137 and am-241. Bulletin of NNC RK. Issue 3.
6. E.Z. Akhmetov. (2002). Features of presence of caesium-137. Bulletin of NNC RK. Issue 3.

EARLY WARNING ENVIRONMENTAL RADIATION MONITORING SYSTEM

NECATI KÜÇÜKARSLAN, ADEM ERDOĞAN, AHMET GÜVEN, YUSUF GÜLAY
Turkish Atomic Energy Authority
Çekmece Nuclear Research and Training Center-İstanbul
*Emergency Response Center-Ankara, TURKEY.

Corresponding author: necati@nukleer.gov.tr

ABSTRACT:

In recent years, especially after the Chernobyl Accident, automated radiation monitoring systems increase its importance in the field of radiation protection. To realize such a system, Turkish Atomic Energy Authority, Çekmece Nuclear Research and Training Center initiated the "Early Warning Environmental Radiation Monitoring System Design and Development Project" in the year 1995. Developed system presently is operating successfully in 67 sites in Turkey.

Keywords: environment, radiation monitor, monitoring, early warning, background, RESA.

INTRODUCTION:

After the Chernobyl Accident, automated radiation monitoring systems increase its importance in the field of radiation protection [1,2,3,4]. To realize such a system, Turkish Atomic Energy Authority (TAEC), Çekmece Nuclear Research and Training Center initiated the "Early Warning Environmental Radiation Monitoring System Design and Development Project" in the year 1995.A radiation-monitoring system has been locally developed that continuously observes the environment for radiation activity. The system consists of intelligent remote probes and a server computer that communicates over public telephone lines. Special detector software has been written that handles the data obtained from the Geiger-Muller tubes. Measurement results are stored on the remote station until transferred to the server computer for later retrieval or analysis. The system has been designed to minimize human interaction by means of programmable server software, which controls routine or periodic tasks such as retrieving data and maintaining detectors. In case of an abnormality in the detected radiation activity, the system automatically raises visual and audio alarms. The server software provides the user with an easy to use interface based on graphical data presentation [5]. The monitoring system is in operation for 3 years with 67 remote stations located all around Turkey.

PROCEDURE:

Early Warning Environmental Radiation Monitoring System (RESA) consists of two main parts. REMOTE STATIONS for radiation detection and RESA CENTER with personal computer running special software called RESA Server. The Center communicates with the remote stations through telephone lines.

M.K. Zaidi and I. Mustafaev (eds.), Radiation Safety Problems in the Caspian Region, 33-41.
© 2004 *Kluwer Academic Publishers. Printed in the Netherlands.*

The REMOTE STATION consists of a "smart probe" and an uninterruptible power supply unit (UPS) containing a modem.

The micro controller 87C51FA from the MCS-51 family is used in the probe.

The probes continuously measure radiation dose rate from background to 400 R/h by two Geiger Muller (GM) detectors sensitive to gamma and X-rays. If one of the detectors is defect the measurement is continued with the other detector. The pulses coming from the detectors are counted separately during one minute. Dead time corrections are made for these counts and radiation dose rates are calculated in $\propto$R/h. There are two operating modes that can be selected from the center:

a. Fixed precision mode: Measurement in the fixed precision mode continues until the measurement reaches 95% statistical accuracy. The measuring period is proportional to the radiation dose rate.

b. Average mode: The measurement in average mode is terminated at the end of the time interval determined by the center which can be from 1 to 240 minutes. In this mode measurement accuracy is proportional to the radiation level.

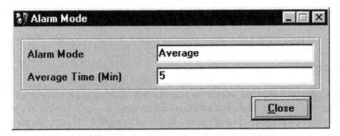

If the alarm level is exceeded, the probe changes its operation mode to the alarm mode and continues operation in this mode. The alarm mode can be selected as "fixed precision" or "average" measurement mode. At the termination of alarm mode, if the measurement result does not exceed the alarm level the probe returns to its normal operating mode, but if the measured value exceeds the alarm level the center is informed with the alarm condition.

Operating in these multi modes allows a comprehensive database formation of the country's radiation map besides the warning capability of the system.

At the end of the measurement, the result and other necessary information are written as a "record" in the probe's memory. Maximum 2032 records can be written in the probe's memory.

The record consists of the following information:

- Date and time of the measurement.
- Measurement result (in $\propto$R/h).
- Measurement mode (fixed or average).
- Measurement duration.
- Alarm condition.
- Alarm level.
- Precision (in %).
- Confirmation of the result with both detectors (in case of alarm state).
- Record memory %70 full.
- Battery condition.
- GM1 detector condition.
- GM2 detector condition.
- Occurrence of the power-off.
- Occurrence of the program restart (watchdog occurrence).
- Sum check of the record bytes.

After each measurement the alarm condition and the status of the probe are tested and it is decided whether the center will be informed or not.

A condition of the probe's searching the center:

- The alarm condition.
- GM1 defect.
- GM2 defect.
- Low battery.
- Record memory at least 70 % full.
- Power-off occurrence.
- Watchdog occurrence.

Probe's identification number, password, telephone number and calibration parameters are stored in the non-volatile memory. Measurement and communication programs are running in the probe simultaneously to provide continuous measurement. The Uninterruptible Power Supply (UPS) unit supplies the probe and the modem. At mains failure the probe continues operation through the accumulator during 10 days. When the mains voltage restores the accumulator will be charged automatically.

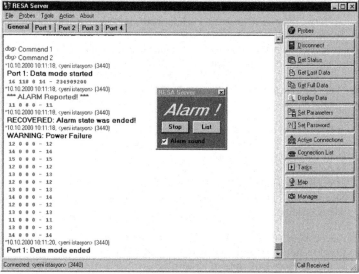

This is the main page of the RESA SERVER.

The server software has been developed with the PASCAL programming language to run under the Microsoft Windows operating system. It may be used for remote data querying. In this mode the program is a read-only browser and does not allow any operations such as changing parameters and connecting to probes. This property provides any user with proper password to connect to the server computer and query radiation information remotely. The server software has been designed to be multi-lingual for international usage and the current version supports Turkish and English. The server software manages the stations and the data obtained from measurements to minimize human interaction to the system.

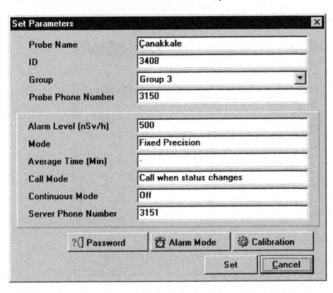

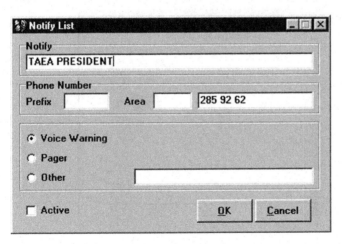

When extra telephone lines are provided as many as the number of serial ports of the computer the server communicates with more than one station simultaneously. The user of the server program may query and change probe parameters used to communicate and measure radiation.

If a radiation alarm is informed by the remote station, the server reports this condition to the user or authorized personal even at home over telephone lines by audio and visual warning.

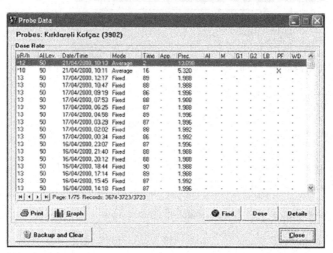

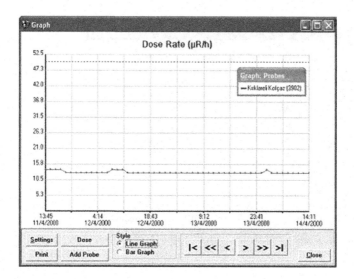

Data accumulated in the databases of the server may be displayed in tabular form. Probe data stored in the databases may be printed, transferred to text files or exported to Microsoft Excel for later processing.

The software has been designed to provide the user with a graphical environment as date time versus dose rate. To compare data of especially neighbor station the user may display overlaid graphics. Each probe is presented with a different color.

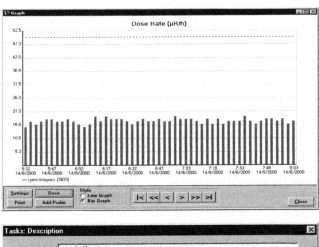

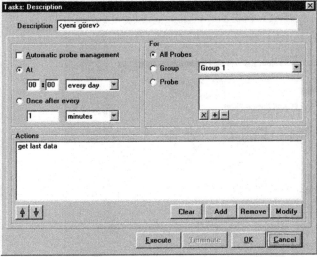

To display unscaled graphics with respect to time the user may choose the Bar Graph form. Dose calculation is possible with selected time interval.

A task module can be programmed by the user to realize all operations related remote stations single or periodically.

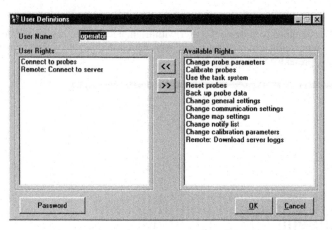

A security system has been implemented to protect sensitive data. Each user has a user name and password for specific operations. The user rights that may be specified by the administration for a user are listed in user definition form shown on the screen.

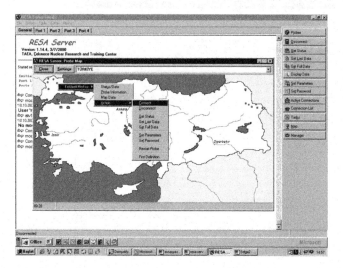

All the operations on the remote stations may be activated through maps. So the user can locate these stations easily.

CONCLUSION:

The main goal of RESA is to have prompt information prior/during an accident. In addition to that RESA gave us valuable information about the background levels for the selected areas of our country.

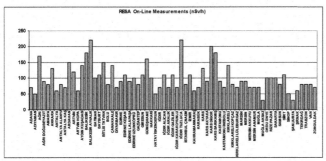

Background radiation level obtained from the RESA stations installed around of Turkey.

REFERENCES:

1.   IAEA. (1996). International Basic Safety Standards for Protection Against Ionizing Radiation and for the Safety of Radiation Sources. Safety Series:115.
2.   Euratom Council Directive 96/29. (1996). Basic safety standards for the protection of the health of workers and the general public against the dangers arising from ionising radiations.
3.   ICRP. (1990). Recommendations of the International Commission on Radiological Protection, Report No. 60.
4.   R.M. Alexakhin Radiation Protection of Humans and Biota in the Environment (www.oita-nhs.ac.jp/~irpa10/CD-ROM/Full/01279.pdf).
5.   P.H. Jensen, Riso Nat.Lab., Denmark.

URANIUM PRODUCTION AS A BYPRODUCT FROM YARIMCA PHOSPHORIC ACID PLANT

G. ÖNAL and S. ATAK

Istanbul Technical University, Mining Faculty, Mineral and Coal Processing Section, 34469, Maslak, Istanbul, TURKEY.

Corresponding author: onalg@itu.edu.tr

ABSTRACT:

Uranium production from the phosphoric acid products of Yarımca Fertilizer Plant is under study. After examination of the phosphate rocks and the acid products, solvent extraction tests were conducted to determine the effects of acid concentration, solvent concentration in kerosene, contact time and acid solvent ratio on the recoveries of uranium. Following the extraction tests, acidic and basic stripping were applied to organic phase and uranium was precipitated as yellow cake from the stripping solutions. In the stripping tests mainly aqueous and organic phase ratio and the stripping time were investigated.

Keywords: Uranium, fertilizer, phosphoric acid and kerosene.

INTRODUCTION:

Uranium recovered from wet process phosphoric acid accounts for about 5 percent of current world production. About 2000 tons of uranium is produced each year as a byproduct from the several phosphoric acid plants located in the USA, UK, Canada and Spain [1, 2].

Yarimca phosphoric acid plant, which is subjected to this study, consists of phosphoric acid, super-phosphate and triple super-phosphate units. Phosphoric acids with 28 and 50 percent P_2O_5 contents are produced in this plant with the annual capacity of 150.000 tons, on the basis of the acid containing 50 percent P_2O_5. In this paper, the properties of the raw materials, which are fed to Yarimca phosphoric acid plant, were investigated as well as the phosphoric acid products. Extraction tests were conducted to estimate the best extraction conditions. After applying 5-stage extraction under the optimum conditions, the stripping tests were conducted and uranium was recovered as yellow cake after precipitation.

MATERIALS and METHOD:

The properties of phosphate rocks from Morocco, Senegal and Togo were fed to Yarimca phosphoric acid plant during this investigation, are shown in Table 1.

Dilute and concentrated phosphoric acid was produced from the phosphate rocks in the plant. The properties of these products can be seen at Table 2. Concentrated and dilute phosphoric

M.K. Zaidi and I. Mustafaev (eds.), Radiation Safety Problems in the Caspian Region, 43-50.
© 2004 *Kluwer Academic Publishers. Printed in the Netherlands.*

acids contain respectively 142-260 and 68-120 mg/l U_3O_8, with the highest value in the acid produced from Morocco phosphate rocks. Uranium recoveries of the phosphoric acids were calculated as 80-95 percent metal, after uranium analysis of tailings, which contain not more than 6 ppm U_3O_8.

Table 1. Properties of the phosphate rocks.

Element	Morocco	Senegal	Togo
U_3O_8 ppm	136	85	95
P_2O_5 %	33.01	36.72	36.75
CaO %	53.53	52.40	50.12
SiO_2 %	1.09	2.80	2.20
Al_2O_3 %	0.36	1.04	1.02
Fe_2O_3 %	0.22	0.19	1.05
MgO %	0.49	0.13	0.13
K_2O %	0.05	0.01	0.04
Na_2O %	0.18	0.04	0.30
F %	3.30	3.35	3.50
Cl %	0.08	0.05	0.15
SO_3 %	1.85	0.32	0.57
Loss of Ignition	6.90	2.91	4.12

Table 2. Properties of the phosphoric acids.

Analysis	Morocco		Togo		Senegal	
	Dilute	Conc.	Dilute	Conc.	Dilute	Conc.
U_3O_8 mg/l	120.0	260.0	68.0	142.0	100.0	215.0
P_2O_5 g/l	332.5	850.4	329.4	808.3	337.8	755.8
MgO g/l	4.4	10.8	0.7	1.5	0.8	1.9
K_2O g/l	0.3	0.4	0.6	0.8	0.3	0.3
Na_2O g/l	1.5	2.3	1.4	1.9	1.0	0.2
CaO g/l	0.2	---	1.2	---	0.4	0.2
SO_3 g/l	24.9	47.9	18.6	34.1	12.9	18.4
Fe g/l	1.8	4.4	6.1	9.3	5.4	11.6
Cl g/l	Trace	Trace	Trace	Trace	Trace	Trace
F g/l	13.2	25.60	12.5	20.8	11.83	19.70
Color	Green	D. Green	Brown	D.Brown	Green	D. Green
P_2O_5 %	26.2	53.2	26.1	50.5	26.6	47.8
Sp. Gr.	1.27	1.60	1.26	1.60	1.27	1.58

EXPERIMENT:

Solvent Extraction

Solvent extraction tests were applied to dilute and concentrated phosphoric acids containing 120 and 260 mg/l U_3O_8, respectively. Organic solvent were chosen as octyl pyro-phosphoric acid

(OPPA) diluted in kerosene [3]. After determination of the oxidation potentials of acids [4], which were sufficient for the extraction studies, experiments were conducted to investigate the optimum conditions relating to the OPPA concentrations in kerosene, phosphoric acid concentrations, OPPA-acid contact time and the ratio of OPPA and acid. At the end of these extraction tests, 5-stage extraction was applied to dilute phosphoric acid under the optimum conditions [5,8,11].

Experimental Procedures

A 240 ml separation funnel was used in the extraction tests with a glass stirrer in it. Octyl pyrophosphoric acid was prepared in the laboratory by mixing octyl alcohol and phosphor pentaoxide [9]. This organic solvent was diluted in kerosene (1-7 percent) before using in the extraction tests. Experiments were conducted at room temperature according to the results of the previous tests [6,10].

OPPA Concentration in Kerosene

In the first series of experiments; OPPA concentration in kerosene was changed between 1 and 6. Experiments were conducted under the constant conditions of 1:1 OPPA-acid ratio and two minute contact time for dilute and concentrated acids. The effect of OPPA concentration to extraction recoveries of uranium are shown in Figure 1. Extraction recoveries increase linearly until 4 percent in both acids with higher values for the dilute one. Recoveries are 50 and 75 percent respectively; at the point of 4 percent OPPA in kerosene for concentrated and dilute acids.

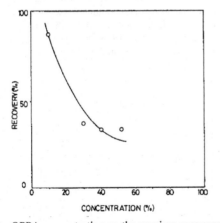

Figure 1. Effect of the OPPA concentration on the uranium recovery

Effect of the Phosphoric Acid Concentration:

Experiments were conducted to determine the proper acid concentration for uranium extraction from the phosphoric acids. In these experiments, the constant conditions were OPPA concentration in kerosene (2 %), OPPA-acid ratio (1:1) and the contact time (2 min). As shown in Figure 2, uranium can be recovered much better in dilute acid than the concentrated one with the highest value of 90 percent uranium at 10 percent of acid concentration.

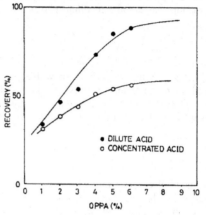

Figure 2. Effect of the phosphoric acid concentration on the uranium recovery

Effect of the Contact Time

The contact time between the extractant and the aqueous solution was investigated for dilute and concentrated phosphoric acids. In this series of experiments, the time was changed between ½ and 10 minutes using two percent OPPA in kerosene and 1:1 OPPA-acid ratio. The recoveries of uranium extraction remained the same after the first minute (Figure 3), with the values of 50 percent in dilute acid and 40 percent in the concentrated one.

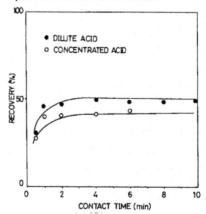

Figure 3. Effect of contact time on the uranium recovery

Effect of OPPA-Acid Ratio

In the last series of extraction tests, the effect of OPPA-acid ratio were investigated between the values of 1:16 and 1:1. During this group of experiments, OPPA concentration and contact time were constant as 2 percent and one minute, respectively. As shown in Figure 4, the extraction recoveries slightly increase (between 30 and 40 percent) in concentrated acid with the increasing of the OPPA-acid ratio, although these values have changed between nearly 10 and 50 percent for dilute acid.

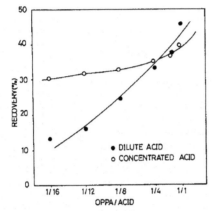

Figure 4. Effect of the OPPA-acid ratio on the uranium recovery

Stage Extraction Tests

Five stage extraction was applied to dilute phosphoric acid in order to recover the most of uranium in the organic phase. Experiment was conducted under the following conditions:

OPPA Concentration: 5 percent in kerosene
Contact Time: 1 minute

OPPA-Acid Ratio: 1:4
Stage Number: 5 (fresh extractant was added in each stage)

The diagram for 5 stage extraction can be seen in Figure 5. Uranium recovery in the organic phase is only 46 percent after the first stage. This value can be raised to 67, 81, 94 and 98 percent by applying 4 more stages using fresh organic solution at each stage. As a result, after five stages of extraction, only 2 percent of uranium can be left in the dilute phosphoric acid.

Stripping and Precipitation of Uranium

Strong acids and bases are recommended [7] to strip uranium from the alkyl pyro- phosphoric acid type solvent. In this study, hydrochloric acid (HCL) and sodium carbonate (Na_2CO_3) are chosen as tripping agents depending on the results of the previous study [9]. Stripping tests with 10 N HCl solution were conducted after oxidation of U^{+4} to U^{+6} by 4 g/l sodium chlorate. In the case of sodium carbonate, no oxidation agents were used for this purpose. The stripping time, stripping solution-OPPA ratio were investigated in the experimental study and stage stripping was applied with both reagents under the optimum conditions. At the end of this study, uranium was recovered as yellow cake from the aqueous solutions.

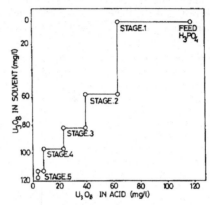

Figure 5. Effect of stage extraction on the uranium recovery

Effect of Stripping Time:

The stripping time was changed between 0.5-60 minutes in the experiments with both 10 N HCl and 1 M Na_2CO_3 solutions under the conditions of 2 percent OPPA in kerosene containing 0.5 g/l U_3O_8 and the ratio of 1:1 between stripping agent and OPPA. As shown in Table 3, two minutes of stripping time is sufficient for Na_2CO_3 solution but more than 30 min. are necessary for appreciable uranium recovery with HCl solution.

Table 3. Effect of stripping time on the uranium recovery

Time	U Recovery %	
Mm	10 N HCl	1 M Na_2CO_3
0.5	45.3	98.0
1.0	65.0	99.2
2.0	71.1	100.0
8.0	75.4	100.0
15.0	78.4	---
30.0	88.2	---
60.0	99.0	---

Effect of Stripping Agent-OPPA Ratio:

Under the same conditions of the previous experiments applying 2 minutes of the stripping time changed the ratio of aqueous and organic phase. Table 4 presents stripping recoveries of uranium depending on the acid-OPPA and carbonate-OPPA ratio. Although a small amount of aqueous solution is sufficient in basic stripping, acidic stripping requires almost equal amounts of organic and stripping solutions.

Stage Stripping Tests:

Stage stripping tests were performed with 1:2 and 1:4 acid-OPPA ratios and 1:10 carbonate-OPPA ratio under the previous conditions. It can be seen in Table 5 that 4 stage stripping for 1:2 ratio and 7 stage for 1:4 ratio is necessary when HCl is used as stripping agent. In the case of Na_2CO_3, two stage stripping is sufficient for more than 99 percent of uranium recovery.

Table 4. Effect of the aqueous and organic solution ratio on stripping of uranium.

| Ratio | U Recovery % | |
Aqueous/OPPA	10 N HCl	1 M Na_2CO_3
1:1	71.1	99.2
1:2	45.8	98.3
1:4	13.0	96.1
1:8	---	93.2
1:10	---	92.4

Table 5. Effect of stage stripping on the uranium recovery

| Stage No | U RECOVERY % | | |
	Acid: OPPA 1 : 2	Acid: OPPA 1 : 4	Carbonate: OPPA 1 : 10
1	45.8	23.0	92.4
2	72.9	44.9	99.5
3	89.6	68.3	100.0
4	99.0	80.7	---
5	---	93.7	---
6	---	98.4	---
7	---	99.7	---

Production and Properties of Yellow Cake

Uranium was precipitated from acidic solution by neutralization with ammonia. After filtration and drying at 105°C, the yellow cakes containing 13-18.4 percent U_3O_8 were produced with 99.5 percent recovery from the acidic solutions containing 2 and 3 g/l U_3O_8.

In the case of Na_2CO_3 solutions, at first pH value was dropped to 3 with sulphuric acid. Following this treatment uranium was precipitated again by neutralization of this acidic solution with ammonia. Yellow cakes containing 30-46.5 percent U_3O_8 were produced with the recoveries of 97.8 percent depending on the concentrations of stripping solutions which were changed between 7.2 and 20 g/l.

CONCLUSIONS:

The imported phosphate rocks consumed in Yarimca Phosphoric Acid Plant contain 85-136 ppm U_3O_8. 80-95 percent of uranium in these rocks can be recovered in the acid phase after phosphoric acid manufacturing. Dilute and concentrated phosphoric acids are produced at Yarimca Phosphoric Acid Plant, containing 26-27 percent and 48-53 percent P_2O_5 respectively. Uranium values of these acids are 70-120 mg/l U_3O_8 in the dilute and 140-260 mg/l U_3O_8 in the concentrated one. Dilute phosphoric acids are most suitable for the solvent extraction process with octyl pyro-phosphoric acid as extractant. The best conditions for extraction were found as 5 percent solvent in kerosene, 1:4 ratios between organic and aqueous phase and under this condition, 98 percent of uranium could be extracted after 5-stage extraction. Stripping of uranium from organic phase with HCl requires high concentration of acid (10 N), high aqueous: organic ratio (1:1-1:4) and long stripping time. Four stages stripping for 1:2 ratio and 7 stage stripping for 1:4 ratio are necessary for high uranium recovery. Sodium carbonate as stripping agent provides appreciable results at short stripping time and low aqueous-organic ratio (1:8-1:10) when its concentration in water is 1 mol/l. Yellow cakes containing 13-18.4 percent U_3O_8 are produced from acidic solutions after neutralization with ammonia. In the case of basic solutions 30-46.5 percent U_3O_8 in yellow cakes were obtained depending on the concentration of the stripping solutions.

REFERENCES:

1. Uranium Extraction Technology. (1983). OECD, Paris.
2. N. Birsen. (1987). Dünya Uranyum Kaynakları, Yeterliliği ve Türkiye'nin Uranyum Potansiyeli, Türkiye Atom Enerjisi Kurumu, Ankara.
3. G. Onal and D. Maytalman. (1978). Yarımca Fosforik Asit Tesisinde Uranyumun Yan Ürün Olarak Kazanılması, TÜBİTAK MAG-G-440, İstanbul.
4. F.J. Hurst, D.J. Crouse. (1974). Recovery of Uranium from Wet-Process Phosphoric Acid by Extraction with Octyl-Phenylphosphoric Acid, Ind. Eng. Chem. Process Des. Develop.13 (3). 286-291.
5. R.R. Gristead, K.G. Shaw and R.S. Long. (1955). Solvent Extraction of Uranium from Acid Leach Slurries and Solutions, Int. Cong. On Peaceful Uses of Atomic. Energy, Geneva, 8:71-76.
6. N.P. Galkin, B.N. Sudorikov. (1964). Technology of Uranium, Atomizdat, Moskova 170-200.
7. K.M. Rafajfeth, A. Kh. Al-Matar. (2000). Hydrometallurgy 56. 309.
8. OECD, Environmental Activities in Uranium Mining and milling. (1999). Report by OECD-Nuclear Energy Agency and IAEA. 29
9. A. Dahdouh, H. Shlewit, S. Khorfan, Y. Koudsi J. (1997). Radioanal. Nucl. Chem. 221, 183.
10. L.C. Scholten, C.W.M. Timmermans. (1996). Fertilizer Res. 43.
11. F.J. Hurst. (1989). The Recovery of Uranium from Phosphoric Acid, International Atomic Energy Authority, Vienna, TEC DOC. 533.

BIOTESTING OF RADIATION POLLUTIONS GENOTOXICITY WITH THE PLANTS BIOASSAYS

N. KUTSOKON, N. RASHYDOV, V. BEREZHNA, D. GRODZINSKY
Institute of Cell Biology and Genetic Engineering
National Academy of Sciences of Ukraine
Kiev, UKRAINE.

Corresponding author: nrashydov@yahoo.com

ABSTRACT:

Using the Tradescantia and Allium-assays, we analyzed the changes induced both on genes and chromosomes, evaluate the genetic damage of radiation pollutants, studied gene mutations in Tradescantia-SH. Both assays used are supposed to be the sensitive biomonitors of the genotoxicity of environmental factors. Our results demonstrate contaminated soils effects both on level of mutations and frequency of morphological abnormalities in Tradescantia. We also determined the characteristic feature of gamma-irradiation in high doses in Allium-assays.

Keywords: Genotoxicity, gamma-irradiation, ^{137}Cs and [241]Am-pollution, Tradescantia and Allium-bioassays, chromosome aberrations.

INTRODUCTION:

 It is well known that radiation effects on the genome of all organisms. Being cheap and easy for use the plant test-systems could be used to predict the genetic damage of different pollutants. The purpose of this work was to evaluate the genetic damage of radiation pollutants with the plant assays Tradescantia and Allium. These assays let to analyze the changes induced both on genes and chromosome levels.

PROCEDURE:

The plants of Tradescantia clone 02 (Figure 1) were grown on soil samples polluted by Chernobyl accident. These samples near Chernobyl and Kopachi regions were taken. The levels of ^{137}Cs and [241]Am in Kopachi sample were approximately ten times as big as those in Chernobyl sample (20.8±0.1 kBq/kg for radionuclide ^{137}Cs and 87.0 ± 7,0 kBq/kg for radionuclide [241]Am versus 1.1 ± 0.1 kBq/kg and 13.0 ± 2.0 kBq/kg accordingly). Control plants were grown on samples where levels of these radionuclides were too low. [137]Cs polluted specially soil as a positive control was used. To study the possibility of adaptation the plants were grown during 3 periods of flowering through 8 – 9 months at the same soil samples. After recovery period (7 days) flowers were analyzed at the morning time daily during flowering [1]. Parameters analyzed were frequency of pink mutations in stamen hairs cells and morphological abnormalities include

M.K. Zaidi and I. Mustafaev (eds.), Radiation Safety Problems in the Caspian Region, 51-56.
© 2004 *Kluwer Academic Publishers. Printed in the Netherlands.*

stamen union and deficiency, quantitative and morphologic alterations of anthers, petals and sepals, morphologic alterations of pistils.

The clastogenic effects of γ-irradiation in Allium-assay on induction of chromosome aberrations in model experiment were estimated. Air-dry seeds of Allium cepa L. were gamma-irradiated in dose range 1 – 40 and 50 – 300 Gy using arrangement with the ^{60}Co isotopes. Cytogenetic effects were studying in the root tip cells test-system [2,3]. The frequency of aberrant anaphases, average number of aberrations on aberrant cell and the intercellular distribution of aberrations were determined. All results were statistically processed, comparison between the experimental variants and controls were conducted by χ^2-method and t-test.

RESULTS AND DISCUSSION:

The level of pink mutation events in Tradescantia-SH induced by soil samples from Chernobyl was lower then that induced by sample from Kopachi, but both levels were statistically higher then compared with control (Table 1). In addition to high level of gene mutations the plants that were grown on most polluted soils samples from Kopachi demonstrated the morphological abnormalities such as stamen union and alteration, flowers underdevelopment (Figure 2).

Our results demonstrate radionuclide contaminated soils effects both on level of mutation events and frequency of morphological abnormalities in Tradescantia clone 02. Positive ^{137}Cs control didn't effect on the morphological abnormalities rate. We supposed the [241]Am soil pollution was more effective then compared with ^{137}Cs pollution.

Table 1: Mutation events in stamen hairs and morphological abnormalities induced by radioactive polluted soil in Tradescantia clone 02

Soil sample	Flowering period	Dose, cGy	Mutation events on 1000 stamen hairs	Frequency of morphological abnormalities, %
Control	1	0.07	0.3 ± 0.2	2.2 ± 2.0
Control	2	0.07	0.3 ± 0.1	3.3 ± 1.0
Control	3	0.07	0.2 ± 0.1	2.0 ± 1.0
Control [137]Cs	3	0.28	4.3 ± 0.5	2.8 ± 1.4
Chernobyl	1	0.86	3.0 ± 0.8	6.8 ± 2.1
Chernobyl	2	0.86	6.1 ± 0.4	8.6 ± 1.5
Chernobyl	3	0.86	4.5 ± 0.4	8.0 ± 1.5
Kopachi	1	17.30	4.8 ± 0.4	8.7 ± 1.5
Kopachi	2	17.30	8.4 ± 0.3	12.2 ± 1.2
Kopachi	3	17.30	7.9 ± 0.6	12.0 ± 1.2

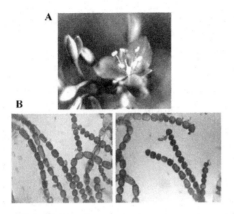

FIGURE 1 TRADESCANTIA, CLONE 02. A – FLOWER; B – STAMEN HAIR CELLS.

Figure 2. Morphological abnormalities in Tradescantia, induced by irradiation:
1 - stamen union; 2 - quantitative and morphologic alterations of anthers ; 3 - stamen deficiency.

Cytogenetic experiments demonstrate strong clastogenic effects of gamma-irradiation in Allium-test in studied dose range. Results obtained are present at the Figure 3 and 4. Statistically reliable increase of chromosome aberrations frequency was shown when dose of γ-irradiation risen to 5 Gy. Effects were intensified gradually when dose increased and obtained nearly 100 % when doses were higher then 200 Gy.

The strong effects on the average number of aberrations on aberrant cell were shown when the dose was 50 Gy and higher, and were less obvious in dose range 1 - 40 Gy. In last case results weren't statistically different when compared with the control. Thus, high doses ᴅ-irradiation induced more aberrations in aberrant cells. When the dose was 40 Gy and higher the "rogue" cells were detected (Figure 5) and theirs' number gradually increased with dose range increasing.

A lot of cells with unidentified plural aberrations among the cells with aberrations were under γ-irradiation in doses 150 Gy and higher (Figure 6).

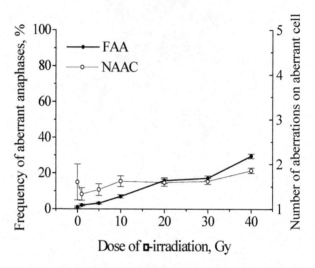

Figure 3. Effects of γ–irradiation in dose 0 and 1 - 40 Gy on frequency of aberrant anaphases (FAA) and average number of aberrations on aberrant cell (NAAC) in Allium cepa L.

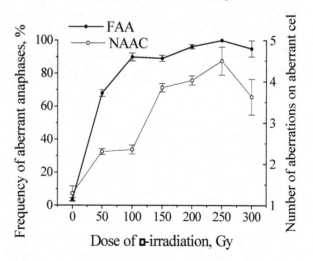

Figure 4. Effects of γ–irradiation in dose 0 and 50 - 300 Gy on frequency of aberrant anaphases (FAA) and average number of aberrations on aberrant cell (NAAC) in Allium cepa L.

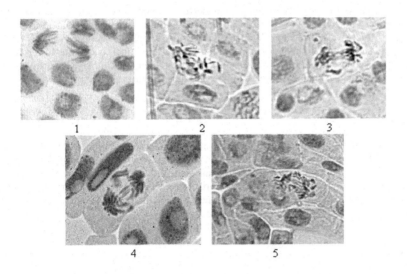

Figure 5. The root tip cells of Allium cepa L. 1 – normal anaphase, 2–5 – cells with plural chromosome aberrations induced by high doses of seeds γ–irradiation.

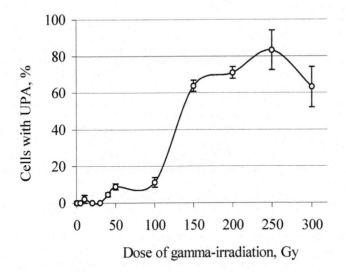

Figure 6. Frequency of cells with unidentified plural aberrations (UPA) (among the cells with aberrations) under γ-irradiation of Allium cepa seeds.

Multiaberrant cells were described among lymphocytes in people cytogenetics studies and their origin is understudied [4, 5, 6], because of clear dose-effect dependencies not always were shown [7, 8]. As our results demonstrate in Allium-assay clear dose-effect dependencies on cell damage were shown only in high doses irradiation [9], and were unclear in doses lower 40 Gy. Therefore, we suppose, the level of rogue cells depends little on dose in range of relatively low doses. This fact could explain the absence of dose-dependencies in population studies.

CONCLUSION:

Thus, then studying gene mutations in Tradescantia-SH induced by polluted soil samples and cytogenetic effects of γ-irradiation in root tip cells of Allium cepa L. we demonstrate the genetic damage of radiation pollution. Both assays used are supposed to be the sensitive biomonitors of the genotoxicity of environmental factors. Furthermore, the character feature of gamma-irradiation in high doses in Allium-assay was induction of the cells with plural chromosome aberrations.

REFERENCES:

1. Ma T.H., Cabrera G.L., Cebulska-Wasilewska A. (1994). Tradescantia stamen hair mutation bioassay. Mut. Res. 310. 211–220.
2. Fiskesjo G. The Allium test - an alternative in environmental studies: the relative toxicity of metal ions. (1988). Mut. Res. 197. 243-260.
3. Rank J., Nielsen M. H. (1993). A modified Allium test as a tool the screening of the genotoxity of complex mixtures. Hereditas. 118. 49–53.
4. Neel J.V., Awa A.A., Kodama Y., Nakano M., Mabuchi K. (1992). "Rogue" lymphocytes among Ukrainians not exposed to radiopactive fall-out from the Chernobyl accident: The possible role of this phenomenon in oncogenesis, teratogenesis, and mutagenesis. Proc. Natl. Acad. Sci. USA. 89. 6973–6977.
5. Bochkov N.P, Katosova L.D. (1994). Analysis of multiaberrant cells in lymphocytes of persons living in different ecological regions. Mut. Res. 323 (1-2). 7-10.
6. Lazutka J.R., Lekevicius R., Dedonyte V., Maciuleviciute-Gervers L., Mierauskiene J., Rudaitiene S., Slapsyte G. (1999). Chromosomal aberrations. Mut. Res. 445. 225–239.
7. Sevan'kaev A.V., Tsyb A.F., Lloyd D.C., Zhloba A.A., Moiseenko V.V., Skrjabin A.M., Climov V.M. (1993). 'Rogue' cells observed in children exposed to radiation from the Chernobyl accident. Int. J. Radiat. Biol. 63 (3). 361-367.
8. Testard I., Ricoul M., Hoffschir F., Flury-Herard A., Dutrillaux B., Fedorenko B., Gerasimenko V., Sabatier L. (1996). Radiation-induced chromosome damage in astronauts' lymphocytes. Int. J. Radiat. Biol. 70 (4). 403–411.
9. Kutsokon N., Rashidov N., Grodzinsky D.M. (2003). Unidentified multiaberrant cells as evidence of γ-irradiation in an Allium-test - Abstracts of Sixth International Symposium and Exhibition on Environmental Contamination in Central and Eastern Europe and the Commonwealth of Independent States, Prague. 142-143.

STRUCTURE OF MET30-SER40 SEGMENT IN N-TERMINUS OF HUMAN TYROSINE HYDROXYLASE TYPE1.

IRADA N. ALIEVA, NARMINA MUSTAFAYEVA, DSHAVANCHIR ALIEV
Baku State University, Z. Khalilov, 23, AZ 1043/1, Baku, Azerbaijan

Corresponding author: iradanur@mail.az

ABSTRACT:

Theoretical conformational analysis was performed to stretch of Met30-Ser40 amino acid residues from the N-terminus of human tyrosine hydroxylase type 1(hTH1). Eight types of stable conformations of the sequence with significantly different values of dihedral angles are possible. Two □-turns of the polypeptide chain in an aqueous environment were revealed on the section Pro32-Ile35 and Phe34-Arg37.

Keywords: Tyrosine hydroxylase; Conformation; Spatial organization; Molecular dynamics simulation.

INTRODUCTION:

Tyrosine hydroxylase (TH) is a rate-limiting enzyme for catecholamine biosynthesis [1,2]. TH consists of a catalytic domain and the N-terminal regulatory domain [3]. The catalytic domain is located at the C-terminal two-thirds of the molecule and binds the substrates (L-tyrosine and molecular oxygen) and the cofactor (6R-tetrahydrobiopterin; 6RBPH4) [4]. In contrast, the important roles controlling the enzyme activity have been assigned to the N-terminal end as the regulatory domain [5,6]. A number of studies have been suggested that the spatial configuration of the N-terminus of Human tyrosine hydroxylase type 1 (hTH1) to be highly flexible and specific stretch of amino acids in this region is supposed to be located at the surface of the overall structure of the enzyme [7]. Very little information on the N-terminus was obtained from an earlier crystallization experiment on the TH molecule because TH deleted up to 155 amino acid residues from the N-terminus was used for the experiment [8]. The speculation that the N-terminus of hTH1, especially amino acid residues 1-40 must have a flexible conformation was made [9], too. However, the reason is not precisely known yet. In addition, the crystal structure of phenylalanine hydroxylase indicates that the amino acid sequence corresponding to the amino acid residues 24-41 of TH is a mobile domain [10]. Taken together, these results suggest that the residues Arg37-Arg38 positioned between Ser31 and Ser40 influence the spatial configuration of the hTH1 regulatory domain, which is relevant to the regulation of feedback inhibition in living mammalian cells [11-13].

In this article we present data that show the conformational flexibility of the amino acid residues 30-40 in the N-terminus of hTH1. The conformational properties of the Arg33, Arg37, Arg38 and phosphorylated Ser31 and Ser40, the main electrically charged amino acids within the

M.K. Zaidi and I. Mustafaev (eds.), Radiation Safety Problems in the Caspian Region, 57-67.
© 2004 *Kluwer Academic Publishers. Printed in the Netherlands.*

stretch of Met30-Ser40 sequence have been studied by the theoretical conformational analysis method.

Computational methods:

1. Molecular mechanics calculations:

Calculations are being carried out on the basis that stable conformations of peptide molecule in solution correspond to local minima of a function (referred to, for brevity, as the "conformational energy", E_{conf}) which is the sum of the potential energy for all intrapeptide interactions and the free energy for all interactions involving the solvent. The most stable conformation of a molecule is then the one, which corresponds to that local minimum of the conformational energy surface, which has the largest statistical weight. In molecular mechanics calculations the conformational potential energy of a molecule is calculated as the sum of independent contributions of nonvalent E_{n-v}, electrostatic E_{el}, torsional interactions E_{tors} and hydrogen bonds E_{nb} formation. The energy of nonvalent interactions has been described by Lennard-Jones potential with the parameters proposed by Momany et al. [14]. A contribution of electrostatic interactions has been taken into account in a monopole approximation according to Coulomb's law, with single atomic charges proposed by Momany et al. [14]. The influence of solvent is included in the dielectric constant (□). The value of this dielectric constant varies from program to program [15-18]. The usual justification for using the distance related dielectric that mimics the effect of solvent in the system. Thus the interaction of the two charges from two atoms in a molecule will be diminished by the presence of solvent molecules. It has not yet been established which value of the dielectric leads to the most realistic results in polypeptide conformation calculations and it is advisable to note the effect of trying different options. Value of the □ is taken to be equal to 10 which according to [19] estimates the electrostatic component of a peptide conformational energy in aqueous medium most satisfactorily. A torsion energy has been calculated using the value of internal rotation barriers given by Momany et al. , i.e. for the backbone chain, U_o(□)=0,8 kJ/mol, U_o(□)=2,5 kJ/mol and U_o(□)=83,7 kJ/mol; for the side chains, U_o(□)=11,7 kJ/mol. Small deviations from planarity of either the cis or trans form, with □□= -10° to +10° are thought to be only slightly unfavorable energetically in most peptide bonds. The energy of intramolecular hydrogen bond N-H...O-C was estimated through a potential approximated by Morse curve with the parameters r_o(H...O) = 1.8Å, n = 3(Å)$^{-1}$ and D = -6,2 kJ/mol [14]. A rigid valence geometry of the amino acid sequence was assumed, namely, the searches were made only on torsion angles. The conformational state of each amino acid residue is conveniently described by backbone □, □ and side chain □$_1$, □$_2$,...dihedral angles. The universal sets of low-energy conformational states were used as starting conformations for all amino acid residues [19].

2. Description and selection of initial conformations:

The conformational state of amino acid residue has been determined by identification X_{ij}^{n}, where X characterizes the form of the main chain of a residue (R,B,L,P), n is the number of a residue in the sequence and subscripts i, j...specify the position of the side chain □$_1$, □$_2$..., respectively, so that i or j = 1 corresponds to the angle □ in the range from 0 to 120°; a value of two corresponds to the angle range from 120 to –120° and three from –120 to 0°. The forms of a

residues denotes the low-energy regions of its backbone dihedral angle φ and ψ on the conformational map: R (φ,ψ = -180 | 0°), B (φ = -180 |0°, ψ = 0 |180°), L (φ,ψ = 0 |180°) and P (φ =0 |180°, ψ = -180 |0°). The combination of backbone form of the residues in a given amino acid sequence will specify the backbone form of a fragment.

All backbone forms of a dipeptide can be classified into two types, referred to as shapes: folded *(f)* and extended *(e)*. The *f*-shape is represented by R-R, R-B, B-L, L-L, L-P, P-L, B-P and P-B forms and *e*-shape by B-B, B-R, L-B, L-R, R-L, R-P, P-L and P-P forms. Forms belonging to a particular shape have an analogues peptide chain contour and a similar mutual arrangement of backbones and side chains and thus should exhibit similar medium-range interaction potentialies. For a tripeptide fragment, all possible backbone forms may be specified by four shapes, i.e. *ff, fe, ef* and *ee*. The nomenclature and conventions adopted are those recommended by IUPAC-IUB [20].

3. Computer experiment:

Met30-Ser40 calculation scheme is shown in Figure 1. The analysis of the spatial structure of sequence has been carried out on the basis of step-by-step calculations. According to [19], the final results found to be independent of the way of polypeptide chain dividing into fragment. The optimization procedure has a several steps. A final conformation obtained at a preliminary step is taken as an initial one for the next step. A procedure for the minimization of the fragments global energy was conducted by the method of conjugate gradients using the program [21]. The minimization of a function of many variables involves the evaluation of its first derivatives with respect to all of the independent variables. These procedures generally lead to stationary points (which can be maxima, minima, or saddle points), so the second derivatives also examined to determine which stationary points are indeed minima. The energy minimization was repeated until the minimal values of the global energy retained constant level.

RESULTS AND DISCUSSION:

The initial conformations for the energy minimization of Met30-Ser40 segment were obtained by combining the calculated lowest energy structures of di-, tri-, tetrapeptide fragments as shown in Fig.1. For each backbone form every possible combination of side chain angles of the residues was taken into account. The side chain dihedral angle values $\chi_1 = \pm60$ and 180° were taken into account for methionine, arginine, serine, phenylalanine, glutamine and isoleucine. Only one of the equally probable conformations ($\chi_{2,3} = \pm60$,180°) , i.e. $\chi_{2,3} = 180°$ was taken into account for methionine, isoleucine and arginine. For glutamine and phenylalanine residues, the value of 90°, which corresponded to the stable state, was used for χ_2 also for methionine, isoleusine and arginine $\chi_4 =180°$ was taken.

According to calculation results 61 low-energy conformations in an aqueous environment with significantly different values of dihedral angles are available for Met30-Ser40 segment. This result indicates that the segment in free state has a large conformational mobility. The conformational energy of the calculated structures is varied in range 0-17 kJ/mol (Table 1). The eight low energy structures are given in Table 2. Table 3 consists of geometrical parameters, i.e.

the values of dihedral angles of two lowest energy conformations, named the global

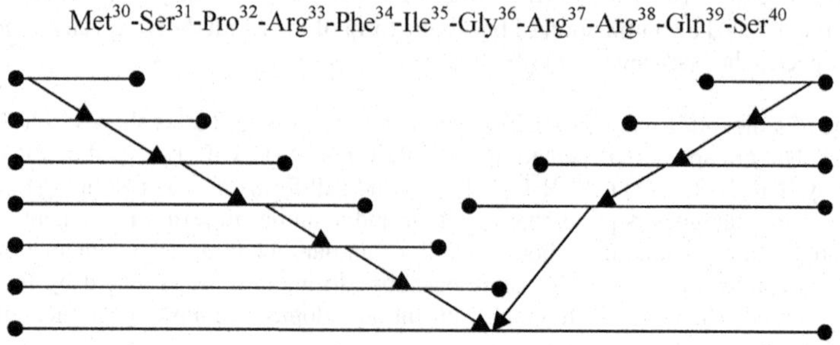

$$Met^{30}-Ser^{31}-Pro^{32}-Arg^{33}-Phe^{34}-Ile^{35}-Gly^{36}-Arg^{37}-Arg^{38}-Gln^{39}-Ser^{40}$$

Figure 1. The calculation scheme of $Met^{30}-Ser^{40}$ segment from N-terminus of hTH1.

conformations. These conformations consist of two reverse turns on the short tetrapeptide fragments Pro32-Ile35 and Phe34-Arg37. The distance between the C^{α}-atoms of Pro32 and Ile35 as well as between the Phe34 and Arg37 is found to approximately 6.7 Å , that

Table 1. The energetic distribution of low-energy conformations of the $Met^{30} - Ser^{40}$ segment

No.	shape	Relative energy (E_{rel}), kJ/mol			
		0-4	4-8	8-12	12-16
1.	eefeffeef	2*	4	2	2
2.	eefffeefef	-	-	-	1
3.	eefffeffef	1	4	3	4
4.	eefffefeef	-	3	7	8
5.	eefffeeefe	-	5	2	5
6.	eefffeeeef	-	-	1	1
7.	eeffffeeef	-	-	-	1
8.	feffffefe	-	-	-	5

* Indicates the number of calculated conformations.

Table 2. The low-energy conformations of $Met^{30}-Ser^{40}$ segment of hTH1.

No.	Shape	Energy contribution , kJ/mol				
		Conformation	E_{conf}	E_{n-v}*	E_{el}	E_{tors}	E_{rel}
1.	eefeffeef	$B_2B_1RB_2R_1B_3PB_1B_2R_{333}R_1$	-140.6	-238.4	78.7	19.2	0.0
2.	eefffeefef	$B_2B_3RR_2R_3B_2LR_3B_2R_{311}R_1$	-136.8	-223.0	68.2	17.8	3.8
3.	eefffeffef	$B_2B_1RR_3R_2B_2RR_3B_2R_{331}R_1$	-136.0	-219.7	69.0	14.2	4.6
4.	eefffefeef	$B_2B_3RR_2R_3B_2RB_1B_2R_{321}B_1$	-135.1	-217.6	66.5	15.5	5.4
5.	eefffeeefe	$B_2B_3RR_2R_3B_2BB_1R_2B_{121}R_1$	-131.0	-218.0	70.7	16.7	10.0
6.	eefffeeeef	$B_2B_3RR_2R_3B_2LB_1B_2R_{321}B_1$	-127.2	-210.4	68.6	14.6	13.8
7.	eeffffeeef	$B_3B_3RR_2L_1B_3LB_1B_2R_{321}B_1$	-125.1	-214.2	72.8	16.7	15.9
8.	feffffefe	$R_3B_3RR_2R_3B_2PB_1R_2B_{121}R_1$	-123.8	-228.4	85.8	19.2	16.7

The energy of hydrogen bonding was included into E_{n-v}

Table 3. The values of the dihedral angels (in degrees) in the low-energy conformations of Met[30]-Ser[40] of hTH1, formed □-turn on the Pro[32]- Ile[35] and Phe[34]-Arg[37] segment . The angle values for each residue are listed in the order □,□,□,□₁,□₂,... (for Pro □ ,□) is characteristic of □-turn. These residues are close to each other and an effective tetrapeptide interaction is found to be contact between them. The results of conformational analysis which reflect the dynamics of the polypeptide backbone are shown in Figure 2, too.

Amino acid residue	Conformation	
	$B_2B_1RB_2R_1B_3PB_1B_2R_{333}R_3$	$R_3B_3RR_2R_3B_2PB_1R_2B_{121}R_1$
Met[30]	-125, 144, 181, 182, 174,	-102, -55, 176, -64, 181, 180,180
Ser[31]	178,180*	-133, 168, 181,-64, 180
Pro[32]	-95, 154, 171, 58, 176	-63, 182
Arg[33]	-46, 174	-105, -53, 173, 195, 179, 181, 180
Phe[34]	-118, 110, 174, 182, 178, 179,179	-133, -62, 178, -67, 103
Ile[35]	-103, -20, 176, 59, 89	-155, 140, 177, 180, 183, 177, 183
Gly[36]	-71, 116, 182, -58, 184, 178, 184	90, -89, 179
Arg[37]	96, -88, 183	-147, 171, 178, 66, 177, 181, 180
Arg[38]	-118, 153, 182, 61, 180, 180, 180	-88, -51, 179, 184, 179, 179
Gln[39]	-116, 120, 181, 179, 181, 180,181	-156, 170, 176, 64, 179, 101
Ser[40]	-93, -54, 181, 59, 65, -83	-81, -43, 180,56, 179
E_{rel}, kJ/mol	-96, 101, 180, 60, 180 0,0	16,0
The hydrogen bonds	Pro[32] C□O ...H-N Ile[35] Arg[33] N□-H...O= C□Ile[35]	Met[30] C□O ...H-N Gly[36] Phe[34] C□O ...H-N Arg[37]

Now, consider the each amino acid conformational properties in detail. The methionine residue at the first position preferably realize the B-form of backbone skeleton. The optimal positions of the Met30 side chain are close to the minima of their torsional potential: □₁ = -60°,180° ($C^{□}$ - $C^{□}$), □₂ = 180° ($C^{□}$ -$C^{□}$) and the deviations by ± 6° from minimal values are possible for all the calculated structures. Specific nonvalent interaction between the Met30 and Phe34 side chains (-16.7 kJ/mol) made an important contribution to the stabilization spatial configuration of the Met30-Ser40 segment. The energy contribution between Met30 and Gln39 in the conformation no.8 (Table 2 and Figure 3) that consists of □-turn segment Pro32-Arg38 is characteristic for all the low-energy structures with *feffffefe* form the polypeptide skeleton. Calculation results indicate that Met30 makes an important contribution towards the dispersion interactions and total conformational energy is varied from -40 to -70 kJ/mol (Figure 3).

The phosphorylation of the serine amino acids at the positions 31 and 40 residing in the N-terminus of TH is one of the most important mechanism involved in the regulation of the catalytic activity of the enzyme. It has been suggested that the TH enzyme inactivated by the binding of dopamine can be reactivated by the phosphorylation of Ser residues in the N-terminus, which might expel dopamine from the TH molecule. In spite of the presence of three phosphorylation sites in the N-terminus of hTH1, i.e. Ser19, Ser31 and Ser40 [22], the fact that only Ser31 and Ser40 are readily phosphorylated to activate hTh1 *in vitro* has become a general consensus [22-25]. A theory that can explain the interrelationship among the positive charge of the specified amino acid residues, phosphorylation of Ser residues and dopamine binding still remains undetermined. So, in this study we analyzed the conformational properties of the Ser31 and Ser40 residues and calculated the energy interactions between the serine residues and other amino acids from the Met30-Ser40 seqment. The conformational dynamics of side chains is estimated quantitatively, too. In all calculated structures Ser 31 backbone chain can realizes only B-form (Table 2) and hydroxyl group is at hydrogen-bonding distance from the carboxyl group of own backbone. Side chain angle $□_1$ for Ser31 can take 60° or -60° in the conformations having reversible bend in the polypeptide chain (Table 3); the angle $□_2$ which defines the orientation of hydroxyl group has not a noticeable conformational mobility. In all calculated structures $□_2$ has a value 180°. It was suggested that Ser31 side chain is strictly fixed in the spatial configuration of hTH1 N-terminus. In the lowest energy conformation $B_2B_1RB_2R_1B_3PB_1B_2R_{333}R_3$ (Table 3) $□$ and $□$ angles can take values -95° and 154°, respectively. The low-energy changes from -95° to –133° of the $□$ angle and from 154° to 168° of the $□$ angle are revealed and the deviations only ± 4° from minimal value for $□_1$ Ser31 are possible for all low-energy structures of the Met30-Ser40 segment. The energy interaction between the Ser31 and other amino acids is varied in the range from -40 to -80 kJ/mol, so, the limitation of conformational flexibility of Ser31 amino acid residue side chain relative the other residues can assumed. Inspite of Ser31, another amino acid, Ser40 is characterized by significant conformational mobility. Its backbone skeleton may realizes B and R-forms and low-energy change in range of -81 to -98 for $□$ and -43 to 101 for $□$ is possible. Calculated results indicate that $□_1$ for Ser40 can takes values 56° and 60° in conformations no.1 and 8 (Table 3), respectively and -60° or 180° in all other structures. The low energy changes ±60° of this angle are allowed. It concludes that such mobility is provided as a possibility for phosphorylation as compared to Ser31. This results correlate with other experimental observations [11].

The presence of a proline residue in the third position of the amino acid sequence significantly reduces its conformational possibilities. Unlike other residues, Pro, because of the rigidly fixed N-C$^□$ bond cannot realize the L and P states. Conformations of the Pro residue depend on only one angle $□$, the diherdral angle for rotation about the central C$^□$ - C$^□$ bond. The values of $□$ 130° (B-form) and -50° (R-form) are allowed . The peptide bond preceding a Pro residue does not have the double bond character that specifies the planar form. Furthermore, only the B and L forms of the peptide skeleton are sterically admissible for the residue preceding proline. It must be noted that conformations with L form of the residue yield energy values approximately 17 kJ/mol higher than those of the B-form, and therefore will not be considered in further investigations. Calculation results indicate that Pro32 involved into formation of the reverse $□$-turn on the tetrapeptide segment Pro32-Ile35 in 80% of all calculated structures.

Arginine residue in position 33 of the amino acid sequence can realizes R and B forms of the backbone and positively charged side chain has a large conformational mobility. In all calculated structures $□$ angle may be changed by ± 15° from its optimal value. Very fixed position corresponding to the value of $□$ angle 110° and -53° in the conformation no.1 and 8, respectively (Table 3) is revealed and their variation lead to increase of conformational energy in the whole segment more than 20 kJ/mol. The rotation of the $\square_2$ and $\square_3$ angles for Arg33 is characterized by

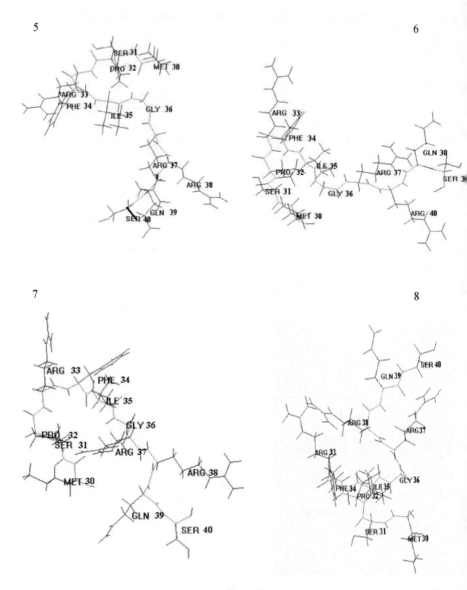

Figure 2 The low-energy conformations of Met[30]-Ser[40] of hTH1, consist of ▢-turn on the Pro[32]-Ile[35] and Phe[34]-Arg[37] segment (the numbering of the conformations corresponds to Table 2)

significant conformational mobility. In all calculated structures $\square_2$ and $\square_3$ may be changed from 150 to 200° and low-energy changes of the $\square_1$ angle in the range 173 -180° are possible. In conformations no.1 and no.5 (Figure 2) Arg33 guanidine group is involved in H-bonding with Phe34 and Ile35 carboxyl groups . One can concludes that Arg33 side chain plays a significant role at formation of additional electrostatic interactions with Phe34 and Gly36 and interacts slightly with Gln39. To clarify how deeply the other arginine residues, Arg37 and Arg38 are involved in the conformation formation of the Met30-Ser40 sequence we study these arginine side chains mobility. The obtained results indicate that unlike Arg33, the other arginine residues are densely packed into spatial configuration of the segment. In the lowest energy conformation these residues form only B-form of their backbone and two values for $\square_1$ = -60 and 60° are possible. So, the conclusion was made that conformational mobility of these arginine backbone is more restricted as compared to preceding arginine residue at position 33. The obtained data allow to conclude that any modification or replacing of Arg37 and/or Arg38 residue may result in dramatic changing in the N-terminus of hTH1. This result is to correlate with the data of previous observations where it is shown that replacement of Arg by electrically neutral and/or negatively charged Glu reduced the inhibitory effect of dopamine on the catalytic activity.

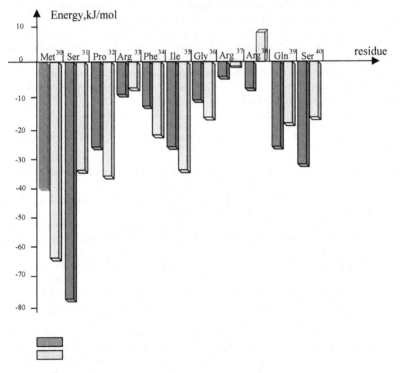

Figure 3. The contribution of energy of interresidues interactions for each amino acid in the conformation of Met^{30}-Ser^{40} with E = 0,0 kJ/mol and E = 16,0 kJ/mol.

The residues Phe34, Ile 35 and Gly36 are characterized by a significant conformational mobility. Calculated results indicate that Phe34 and Gly36 amino acid side chains have a large conformational flexibility. Only value of 60° for $\square_1$ in the Phe34 chain is not allowed. The dihedral angles of backbone and side chains of these residues is varied in the range from -103 to −133° for $\square$ and from −20 to 158° for $\square$ (Phe34); from -71 to −149° for $\square$ and from 114 to 142° for $\square$ (Ile35); from -87 to 96° for $\square$ and from −88 to 104° for $\square$ (Gly36). $\square_2$ angle of Phe34 has a weak conformational mobility, and only low-energy change by $\pm 5^0$ for $\square_2$ is possible. Gly36 can realize the R, B, L and P-forms, i.e. keeps the property of labile residue that is characteristic a lot of know proteins. Analyzing the energy contributions in the low-energy structures (Figure 3) one can come to a conclusion that the glycine residue doesn't significantly contribute to structural stabilization.

CONCLUSION:

Theoretical conformational analysis of the Met30-Ser40 segment in N-terminus of hTH1 indicated that a great number low-energy conformations are allowed. These conformations belong to eight type structures, described by the peptide skeleton shapes as shown in Figure 2. All calculated conformations could make a reversible bond ($\square$ -turn) on the segment Phe34-Arg37. The limitation of conformational flexibility of Ser31, Arg37 and Arg38 amino acid residues is assumed. They are dense packed into spatial configuration of the segment, so any modification or replacing of these residues may result in dramatic changing in the N-terminus of hTH1. A large conformational mobility of the Ser40 amino acid provides a possibility for phosphorylation as compared to Ser31.

REFERENCES:

1. Hufton S.E., Jennings I.G., Cotton R.G. (1995). Structure and function of the aromatic amino acid hydroxylases.Biochem.J. 311, 353
2. Alterio J., Ravassard P.,Haavik J., Biguet N.F., Waksman G., Mallet J. (1998). Human tyrosine hydroxylase isoforms. J.Biol.Chem. 273, 10196
3. Abate, C., Joh, T.H.(1991). Limited proteolysis of rat brain tyrosine hydroxylase defines as N-terminal region required for regulation of cofactor binding and directing substrate specificity. J.Mol.Neurosci. 2, 203
4. Almas, B., Toska, K., Teigen, K., Groehn, V., Pfleiderer, W., Martinez, A., Flatmark, T., Haavik, J. (2000). A kinetic and conformational study on the interaction of tetrahydrobiopteridines with tyrosine hydroxylase. Biochemistry 39,13676
5. Hoeldtke, R., Kaufman, S. (1977). Bovine adrenal tyrosine hydroxylase. J.Biol.Chem. 252, 3160
6. Abate, C., Smith, J.A., Joh, T.H. (1988). Characterization of the catalytic domain of bovine adrenal tyrosine hydroxylase. Biochem.Biophys.Res.Commun.151, 1446
7. McCulloch, R.I., Fitzpatrick, P.F. (1999). Limited proteolisis of TH identifies reduces 33-50 as conformationally sensitive to phosphorilation state and dopamine binding. Arch.Biochem.Biophys. 367, 143

8. Goodwill, K.E., Sabatier, C., Marks, C., Raag, R., Fitzpatrick, P.F., Stevens, R.C. (1997). Crystal structure of TH at 2.3 A and its implications for iherited neurodegenerative diseases. Nat. Struct.Biol. 4, 578

9. Nakashima, A., Mori, K., Nagatsu, T., Ota, A. (1999). Expression of hTH type 1 in E.c. as a protease-deavable fusion protein. Short communication. J.Neural.Transm.106, 819

10. Kobe, B., Jennings, I.G., House, C.M., Michell, B.J., Goodwill, K.E., Santarsiero, B.P., Stevens, R.C., Cotton, R.G., Kemp, B.E. (1999). Structural basis of autoregulation of phenilalanine hydroxylase. Nat.Struct.Biol. 6, 442

11. Nakashima, A., Hayashi, N., Mori, K., Kaneko, Y.S., Nagatsu, T., Ota, A. (2000). Positive charge intrinsic to Arg37-38 is critical for human tyrosine hydroxylase type1. FEBS Lett. 465, 59

12. Nakashima, A., Mori, K., Suzuki, T., Kurita, H., Otani, M., Nagatsu, T., Ota, A. (1999). Dopamine inhibition of human tyrosine hydroxylase1 is controlled by the specific portion in the N-terminus of the enzyme. J.Neurochem. 239, 2910

13. Ota, A., Nakashima, A., Mori, K., Nagatsu, T. (1997). Effects of dopamine on the N-terminus-deleted human tyrosine hydroxylase type1 expressed in Escherichiacoli. Neurosci.Lett.229, 57

14. Momany, F.A., McGuire, R.F., Burgess, A.W., Scheraga, H.A. (1975). Energy parameters in polypeptides. J.Phys.Chem.79, 2361

15. Weiner, S., Kollman, P., Nguyen, D., Case, D. (1986). AMBER Program J.Comput.Chem.7,230

16. Allinger, N., Yuh, Y. (1982). MM2 Program, QCPE 395, Quantum Chemistry Program Exchange, Indiana

17. White, O., Kuddock, J., Edgington, P. (1989). CHEMMIN Program In Computer Aided Molecular Design, W.G.Richard (Ed.), IBC Technical Services

18. Hambly, T. (1991). Inorg.Chem.30,937

19. Popov, E.M. (1979). Quantative approach to conformations.Int.J.Quant.Chem.16, 707

20. IUPAC. (1993).Quantities,Units and Symbols TradeBlackwellScientific,Oxford

21. Maksumov, I.S., Ismailova, L.I., Godjayev, N.M. (1983). J. Struc.Chem.(in Russian) 24, 147

22. Martinez, A., Haavik, J., Flatmark, T., Arrondo, J.L.R., Muga, A. (1996). conformational properties and stability of tyrosine hydroxylase studied by infrared spectroscopy:effect of iron/catecholamine binding and phosphorylation. J.Biol.Chem 271, 19737

23. Haycock, J.W., Wakade, A.R. (1992). Activation and multiple-site phosphorylatin of tyrosine hydroxylase in perfused rat adrenal glands. J.Neurochem.58,57

24. Sutherland, C., Alterio, J., Campbell, D.J., Le Bourdelles, B., Mallet, J., Haavik J.,Cohen P. (1993). Phosphorylation and activationEur.J.Biochem 217.715

25. Ramsey, A.J., Tipzpatric, P.F. (1998). Effects of phosphorylation of serine 40 of tyrosine hydroxylase on binding of catecholamines: evidence for a novel regulatory mechanism. Biochemistry 37. 8980

ECOLOGICAL SITUATION AT "KOSHKAR-ATA" NUCLEAR TEST SITE

K.K. KADYRZHANOV[1], K.A. KUTERBEKOV[1], S.N. LUKASHENKO[1], V.N. GLUSCHENKO[1], M.M. BURKITBAEV[2],
N.G. Kijatkina[1], V. S. Morenko[1], S.P. Pivovarov[1]
[1]Institute of Nuclear Physics (INP), National Nuclear Centre (NNC), Semipalatinsk, Alamaty 480082, [2]Al-Farabi Kazakh National University, Alamaty, Republic of Kazakhstan.

Corresponding author: kuterbekov@inp.kz

ABSTRACT:

Due to particular attention paid by the State, public and international scientific community to the radioecological problems of the former nuclear test sites, these territories are under control and thorough consideration. Much less is paid to radioecological investigations of the uranium mining industry areas. An example illustrating the scale of the problem is tailing KOSHKAR-ATA situated in a close vicinity of Aktau city and Caspian Sea. Initially it was an internal-drainage sedimentation lake; it shallows now. Its total area is 64 km^2. At its naked part of about 10 km^2 the expose dose rate reaches up to 1500 µR/h at some points.

Keywords: Internal Dose, fallout, exposure, population, radioactive waste, and mining.

INTRODUCTION:

Ecological situation in the Mangystau oblast (Kazakhstan) started aggravating in 1960's with exploration of uranium ore, oils, mineral raw and creation of chemical industry in Mangystau without paying due attention to environment issues. According to available data the worsening is connected with a number of ecologically unfavorable problems. A tailing pond KOSHKAR-ATA is the most hazardous place among all objects that makes a atmosphere contaminated with powder radioactive and toxic waste from chemical and mining metallurgic industries. KOSHKAR-ATA represents serious hazard for habitants of Aktau and adjacent inhabited localities [1,2].

The tailing pond KOSHKAR-ATA, representing the drain-free settling pool for industrial, toxic, chemical and radioactive wastes, is 5 km northward to Aktau (Mangystau oblast), which is situated on the shore of the Caspian Sea. The solid sediments of unpurified ordinary domestic drains form a part of the Aktau dwelling region and drained into the tailing pond since 1965. Till 1965 the tailing pond KOSHKAR-ATA represented internal-drained basin with the first mark of bottom - minus 38 m.

Burial of solid radioactive waste of the chemical mining metallurgic plant, where uranium ore was processed, was carried out beyond the control and confidential accountancy in the trench-type burial without hydro-isolation. According to the data of the Mangystau Regional Ecology Department, Aktau, radioactive waste, disposed in the tailing pond is about 360 m.tons with an activity of 11000 Ci.

M.K. Zaidi and I. Mustafaev (eds.), Radiation Safety Problems in the Caspian Region, 69-78.
© 2004 *Kluwer Academic Publishers. Printed in the Netherlands.*

Industrial wastes are represented mainly by phosphogince, phosphomel and pyrite cinders that contain such stable elements as iron, silicon, sulfur, zinc, lead, silver, selenium, cobalt, arsenic, etc., and enhanced concentration of natural radionuclides. Works performed by the Institute of Nuclear Physics in 1999 showed that exposure dose rate (EDR) at shallow zone was 80-150 μR/h. Some places, where EDR was 1500 μR/h and radionuclide contents were up to 548-5000 Bq/kg.

As a result of stable reduction of the water phase level, the area of the naked bottom sediments, that is a source of toxic dust, have increased lately. Under existing hydrogeological conditions of the tailing pond region, there is potential for penetration of liquid waste to aquifers and to the Caspian Sea [3].
Inhabited settlement KORA is situated in western shore of the settling pool. Due to enhanced radiation background, its inhabitants had suffered from continuous enhanced irradiation and poisoning with chemical toxicants. Currently the inhabitants are removed from this inhabited locality.

Thus, continuous and long-term contamination of the land around the tailing pond KOSHKAR-ATA and its plant unambiguously forces to consider regional ecological situation as a critical one. That is why it is very important to take immediate measures to eliminate sources of contamination and restore natural characteristics of environmental situation.

EXPERIMENTAL RESEARCH:

In order to obtain qualitative data on current ecological state of air, water and soil on the territory of toxic and radioactive waste tailing pool KOSHKAR-ATA and vicinity, the field survey were conducted for the year 2003 field expedition. Samples were delivered to INP laboratories and were analyzed and data collected.

Assessment of the main negative factors of the tailing pond KOSHKAR-ATA that influence the environment: since 1997 till 2003, some protective measures had been taken at the site. For example, a trench was constructed along the perimeter to prevent access to the tailing site for motor transport and domestic animals. Surface at the territory adjacent to the effluent disposal line with previously revealed maximal dose rates for ɑ-radiation was covered with 15-20 cm clay soil. Now this territory has developed into a vegetation area. The spring and rainy season had resulted in higher dampness of the surface so there was almost no dusting.

Field investigations were performed in the scope that assures detailed radiometric surveying of shallow parts of the pool and its shore. The sampling of environmental objects were conducted:

Samples of sediment admixtures were taken with exposure 15 days deposited on the pads placed previously; the mapping of current outline was done of the coast-line/water edge of the toxic waste storage; the footpath gamma-surveying was performed on exposed surface (TED was measured in 150 points) and the gamma-field distribution was mapped (Figure 1), the exhalation level for radon was assessed at the shore-exposed surface of the tailing pool. Radon concentration in near-surface layer of air varied within the limits $0 - 24$ Bq/m^3 at background level $0 - 2$ Bq/m^3 (according to measurements in 22 points); the detailed gamma surveying has been done in the vicinity of surface burial of radioactive waste (RW). Investigated area is as

large as 8700 m² (net 2x2 m²). At several points TED is as high as 700 μSv/h (Figure 2); 1ˢᵗ stage
of monitoring measurements has been completed of air aerosols at 10 stations;

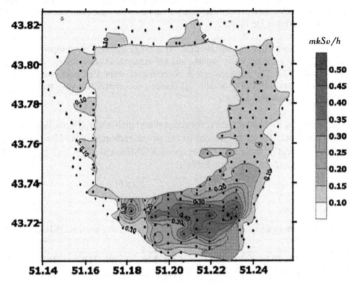

Fig. 1. Map outline of a gamma-field (h=5 cm) in vicinity of the tailing pool KOSHKAR-
ATA.

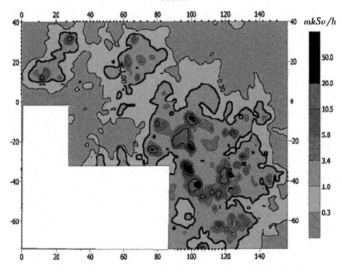

Fig. 2. Map outline of gamma-survey in vicinity of surface burial of RW (h=5 cm).

sampling of layer-by-layer samples has been completed in bore-pits to assess vertical distribution of radionuclides in silt sediments at the shore exposed surface of the tailing pool; and the species identification has been performed and samples of vegetation were taken at the waste storage in places where soil fill had been previously performed; that was done to estimate radionuclide accumulation rate by vegetation and their further wind transfer.

The weather station at Aktau collected information on average monthly air temperature value, relative humidity, wind velocity, amount of precipitations, and information on preliminary wind direction for the period January – June 2003. The meteorological observations were also performed with the help of portable meteorological station directly sampling of aerosol; the meteo-data was analyzed together with data obtained at field investigation.

Laboratory determination of concentration for natural and artificial radionuclides, toxic elements and analysis of the results. To study present forms of the radionuclides. The layer-by-layer soil samples were taken directly from the tailing pool KOSHKAR-ATA and performed gamma-spectrometry; the results are presented in Table 1.

Table 1

Gamma-spectrometric analysis of soil samples taken in the tailing pool KOSHKAR-ATA

Sample code	Pb-210 Bq/kg	±	Th-234 Bq/kg	±	U-235 Bq/kg	±	Ra-226 Bq/kg	±	Pb-212 Bq/kg	±	Bi-214 Bq/kg	±	K-40 Bq/kg	±
KSH 165 A	62.3	9.9	56.5	9.7	7.0	5.4	62	39	14.5	1.7	74.3	4.7	286	20
KSH 165B	40.0	8.6	22.1	5.5	9.4	5.3	n/r		26.4	2.4	30.5	3.3	638	36
KSH 165C	38.7	7.6	28.3	6.4	<8.4		n/r		26.4	2.2	34.5	3.4	728	38
KSH 174 A	93	12	50.2	8.8	8.5	4.7	26	15	13.6	1.6	63.3	4.3	266	18
KSH 174B	41.8	8.6	22.3	5.9	6.6	4.9	<14		33.6	2.3	40.2	3.5	649	34
KSH 174C	55	10	40.2	7.4	<8.2		n/r		27.5	2.2	39.2	3.5	617	33
KSH -48A	267	23	193	21	22.2	7.4	398	105	28.3	2.5	385	10	65.3	6.0
KSH -48B	830	39	529	32	45.7	9.4	708	130	30.8	2.8	653	13	252	19
KSH -48C	639	34	401	28	45.1	9.5	673	127	19.8	2.3	657	13	139	12
KSH -54A	246	20	151	16	14.7	6	261	83	11.0	1.7	190.6	7.3	60.7	5.5

KSH-54B	972	40	479	29	44.5	9.2	787	138	35.7	2.8	539	12	419	26
KSH-54C	1133	43	429	30	64.4	10	692	118	40.3	3.1	664	13	464	28
KSH-51A	746	37	438	27	47.5	8.7	498	97	28.1	2.5	502	12	225	17
KSH-51B	660	34	412	26	45.1	8.6	425	87	31.9	2.7	453	11	250	18
KSH-51C	732	38	466	29	53.7	9.5	622	111	31.6	2.8	630	13	211	16
KSH 122A	81	12	42.3	8.9	<9.8		n/r		11.7	1.6	86.9	5.2	115.7	9.7
KSH 122B	52.0	8.6	29.4	5.7	<7.8		n/r		11.2	1.5	48.5	3.7	223	16
KSH 122C	46.3	8.6	23.1	5.1	<7.7		n/r		10.8	1.5	44.9	3.6	297	20
KSH 150A	81	11	41.1	7.7	3.6	4.4	72	49	12.4	1.6	62.6	4.2	192	14
KSH 150B	40.7	8.1	26.9	6.5	<7.9		n/r		26.9	2.1	30.0	3.1	516	29
KSH 150C	32.1	6.5	20.1	4.9	8.0	4.7	n/r		27.0	2.1	34.4	3.3	609	32

The samples were selected from three levels – 0, 30 and 50 cm. The results shown above illustrate a wide range of concentration of natural radionuclides (NRN). For example, in the point 150 (Sample code KSH-150A, B, C) the concentrations of the determined radionuclides differ slightly from the background levels (30-40 Bq/kg on ^{210}Pb and ^{214}Bi). In other points concentration of ^{210}Pb reach 1000 Bq/kg and 700-800 Bq/kg on ^{226}Ra.

Type of vertical distribution of nuclides on depth is considerably different for different points of sampling. Thus, in the point 150 maximal concentration of NRN are fixed in surface sample, in the point 51 the determined values do not change with the depth and in the points 48 and 54 the level of radioactive contamination of soil grows on increasing the depth of sample selection. Based on the obtained data, for investigation of the present state there were chosen the following eight samples: KSH-165A, KSH-174A, KSH-48A, KSH-48B, KSH-54B, KSH-51A, KSH-122C, KSH-150A.

Sample Preparation: mechanical mixing of 4 initial sample weights with distilled water (at S: L = 1:5) during daytime followed by filtration; then they were left for 30-days accumulation for revealing the water-soluble form of the natural radionuclides. The water extracts were found to contain anions: chlorides, sulphates as well as pH and total mineralization. The results are available in Table 2.

Table 2
Determination of anion composition, total mineralization and pH of sample water extracts

Sample Nos	Concentration, mg/l			PH
	Sulphates	Chlorides	Total mineralization	
KSH-122C	2360	2000	7565	6.95
KSH-150A	2230	1080	5590	7.01
KSH-165A	1900	443	3740	7.80
KSH-174A	4890	6510	19000	8.25

The data on the field investigation were organized in an electronic database and analyzed using GIS-technology. Outline of the sampling points at tailing KOSHKAR-ATA is shown on the map (Figure 3).

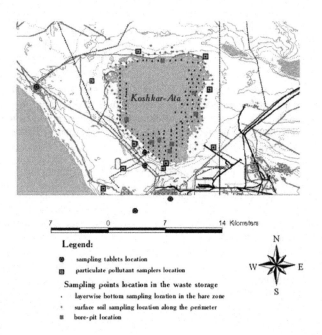

Fig. 3. Outline of the sampling points at tailing KOSHKAR-ATA.

The results showed increased salinity of soils/rocks at the waste storage KOSHKAR-ATA. At KOSHKAR-ATA, there are about 100 observational hydrogeological wells that were set out of operation for a long period of time; most of them can not be used for sampling and require reconstruction. Currently under a task issued by the local (oblast) Mangystau authorities on environmental protection the work on restoration of these wells and underground water monitoring is being performed. The technical information on the hydrogeological wells and underground water samples will be obtained through laboratory investigations. Surface soil sampling was also performed.

Complex assessment of ecological situation at the territories and inhabited localities adjacent to the object KOSHKAR-ATA: The detailed radiometric surveying of inhabited localities was performed: town Aktau, villages Akshukur and Bayandy, Mangystau railway junction was performed and environmental samples were collected.

◻ The measured concentrations of radon in air of living and commercial buildings at Mangystau (28 measurements), v. Akshukur (20 measurements), v. Bayandy (20 measurements) and in Aktau (25 measurements). Radon concentration ranged from 0 to 27 Bq/m^3 for Mangystau, up to 97 Bq/m^3 for Akshukur village, 85 Bq/m^3 for v. Bayandy and up to 27 Bq/m^3 in Aktau.
◻ 14 samples of vegetation were collected in Aktau, Bayandy, and Mangystau.
◻ The surface samples were taken in direction KOSHKAR-ATA – Mangystau.
◻ The air aerosols samples were collected in Aktau, Mangystau, Bayandy, and Akshukur inhabited areas (14 samplings 10 hours each).
◻ For retrospective experimental determination of doses obtained by local population by the method of EPR-dosimetry using teeth enamel there were some sampled teeth removed according to medical prescriptions.

The analysis of the obtained results was performed of over-all radioecological investigation of adjacent inhabited locations with the purpose to assess the influence of the site KOSHKAR-ATA on pollution of the near-surface air. The Microsoft Access database was acquired for storage and processing of attributive information such as sampling point's coordinates, rates of soil and ground water contamination with toxic and radioactive waste. To use the database with analytical results is in progress.

As a result of the field investigation performed in accordance with the schedule, the following samples were taken for further analytical works:

◻ 50 samples of air aerosols in vicinity of the waste storage;
◻ 14 samples of air aerosols in inhabited localities Aktau, Mangystau, Bayandy, Akshukur;
◻ 34 surface soil samples from the tailing perimeter;
◻ 453 layer-by-layer soil samples in 150 points from dry shore part of the tailing;
◻ 121 layer-by-layer from 8 pits of 110-130 cm depth at the dry shore part of the tailing;
◻ 8 surface soil samples in direction KOSHKAR-ATA – v. Bayandy;
◻ 10 surface soil samples in direction KOSHKAR-ATA – v.Akshukur;
◻ 9 surface soil samples in direction of Caspian sea;
◻ 90 layer-by-layer soil samples in 30 points from v.Akshukur;

- 60 layer-by-layer soil samples in 20 points from v. Bayandy;
- 120 layer-by-layer soil samples in 40 points from Mangystau station;
- 54 layer-by-layer soil samples in 18 points from Aktau town perimeter;
- 36 layer-by-layer soil samples in 12 points from Aktau town;
- 72 layer-by-layer soil samples in 24 points from dacha estate site (next to Aktau);
- vegetation samples were taken in Aktau, Bayandy, Mangystau (total 16 samples);
  27 tooth samples of the medically removed teeth.

The field investigation for 2003 has been completed and analyses make it possible to determine the radioecological status of the investigated territories. Radon concentration is considerably below the maximal allowable concentration in the air of living and commercial buildings in inhabited localities.

Meteorological information about the territories adjacent to Aktau town presented by State Enterprise as well as meteorological measurements performed using portable meteorological station evidence for absence (for the period under consideration) of unfavorable factors that might stipulate radionuclide proliferation with wind from the tailing pool territory. At that period wind velocities comprised 2 – 5 m/s and, according to the information on preferable wind directions, prevailed winds were in western, north-western directions. Nevertheless, impact of the tailing pool is possible due to other migration ways available for radionuclides and toxic elements such as migration with surface and ground water, vegetation used as forage.

Since influence of the tailing KOSHKAR-ATA on radioecological situation at adjacent territories by means of dust depends greatly on season changes in soil humidity, monitoring investigations are to be performed at various seasons. The year 2003 is not typical by its meteorological indexes for this territory. Amount of precipitate exceeds the average level. Taking this into account it is supposed to continue monitoring investigation in the years to come to obtain reliable results.

Laboratory determination of concentration of natural and artificial radionuclides, toxic elements and analysis of the results. Within the over-all radioecological investigation of the inhabited localities (Aktau, Mangystau, Akshukur, Bayandy) adjacent to the site KOSHKAR-ATA, gamma-spectrometric measurements were performed of surface soil samples. For laboratory analyses, the samples used were taken at the following sites: Mangystau – 38 samples, Aktau – 54 samples, Bayandy – 20 samples, Akshukur – 30 samples. The analysis was performed of the radionuclide composition of soil samples taken in Aktau, Mangystau and Akshukur to reveal content of ^{137}Cs and natural radionuclides ^{210}Pb, ^{235}U, ^{226}Ra and ^{40}K. The results show that concentration of ^{137}Cs is at the level of global fallout for all the analyzed samples. Concentration of natural radionuclides do not exceed values average for such soil types.

However, after approval of this standard there were introduced additional specifications related to special International experiments "Intercomparison–2" and "Intercomparison–3" performed by IAEA and aimed to verify reliability of EPR-dosimetry data on tooth enamel. NMR laboratory was included into the list of participants of both experiments among other leading laboratories of the world. "Intercomparison – 2" has revealed that considerable error may arise due to peculiarities of preliminary sample preparation of TE. In this connection, sample preparation of TE has been performed in accordance with the results of "Intercomparison – 2".

Experiment "Intercomparison–3" was performed in May–September 2003. It has revealed that, beyond sample preparation, considerable errors may be introduced by individual differences in radiation sensitivity of TE. Taking into account these recommendations, the original plans were changed and it was decided to apply the method of additional irradiation to correctly take into account the differences in individual sensitivity of TE.

Therefore, during the period under consideration a detailed analysis was performed of the results from "Intercomparison–3", fangs of all samples selected for EPR-dosimetry were cut-off; the samples were treated in ultrasound bath with alkaline solution, then dentine was removed with mechanical diamond bore. Remained enamel was transformed into granules of specific size (by means of milling in agathic pounder); their initial EPR-signals were taken. Additional irradiation was performed using the ^{137}Cs source. Participation in the experiment "Intercomparison-3" made it possible to calibrate the dose rate of this installment using a standard IAEA source, so the problem of special calibration was solved.

The method of additional irradiation requires long-term exposures and following detention of a sample after irradiation. To assure precise extrapolation it is required to have not less than 4 additional irradiations with precise doses. During the report period, one irradiation was performed; after detention of the samples there was initiated second registration of their spectra. Determination of absorbed dose by extrapolation to zero will be possible to perform after another irradiations that will tentatively take another one and a half month.

There were performed works on literature review regarding main paths of radioactive and toxic elements proliferation from the territory of KOSHKAR-ATA tailing and on mathematical approaches to simulation of the proliferation processes. The most hazardous ways for proliferation of radioactive and toxic elements are:

◻ wind transportation – can influence territory and air basin of the populated area adjacent to the tailing;

◻ migration with ground water – shore of Caspian Sea located ~5 km from the tailing can be contaminated.

To describe the process of proliferation with wind, i.e. distribution of admixtures (aerosols) in air the stationary model is conventionally used or there can be used a non-stationary model of admixture turbulent diffusion at interface layer. Both approaches imply numerical solution of the equations by methods of mathematical physics at corresponding initial and boundary conditions and source function (inflow of admixture into interface region of atmosphere). Main stages at solving the problem of proliferation with wind for radioactive and toxic elements are the following: choice of the most appropriate model for the case, development of computer codes, typification of atmospheric processes and determination of atmospheric parameters on the basis of known meteorology characteristics of the site, experimental determination of source function, i.e. rate of blowing out of hazardous substances from the naked surface of the tailing KOSHKAR-ATA, making estimation calculations of radioactive and toxic elements proliferation due to wind transfer on the chosen model.

CONCLUSIONS:

The assessment of the ecological situation within the region of the tailing pond KOSHKAR-ATA will include the main following steps:

1. The quantitative data on contamination of the territory of tailing pond of toxic/radioactive wastes KOSHKAR-ATA and adjacent lands with radioactive isotopes and toxic metals, that characterize present ecological situation with air/water/soil in that territory and adjacent lands would be obtained (in particular, in direction towards the Caspian Sea);

2. On the basis of the obtained experimental data, the dose loads on local population throughout studied lands would be determined;

3. Actions aimed to rehabilitation of the shallow zone of the tailing pond would be developed, taking into consideration the local soil/climate conditions;

4. The main elements of rehabilitation actions on the control sites will be created, their efficiency will be estimated;

5. Analysis of potential migration pathways for contaminating and dose-forming elements would be given before and after rehabilitation actions are carried out and forecast of ecological situation evolution would be made;

6. A database on the facility KOSHKAR-ATA is to be developed;

7. The hydrodynamics of underground water at the territory adjoined to the tailing pond and ways of proliferation of the contaminated waste in the direction of Caspian Sea is to be studied with geophysical methods;

8. Recommendations concerning implementation of rehabilitation measures would be developed and submitted to the local executive bodies.

REFERENCES:

1. Kadyrzhanov K.K., Kuterbekov K.A., Akhmetov E.Z., Lukashenko S.N., Dzhazairov-Kakhramanov V. (2000). Radiation-Hazardous Objects at the West and Central Kazakhstan Territory. Presentations - I Eurasia Conference on Nuclear Science and its Application, Turkey. 665 – 673.

2. Kadyrzhanov K.K., Kuterbekov K.A., Lukashenko S.N., Melentiev M.I., Stromov V.M., Shaitarov V.N. (2000, 2002). Overall Examination of the Ecological Situation in the Toxic and Radioactive Wastes Storage "Koshkar-Ata" and Development of Rehabilitation Actions, International Conference on Radiation Legacy of the 20th Century: Environmental Restoration. Proceedings of an Inter. Conference (Radleg-2000). IAEA-TECDOC-1280. IAEA, Vienna, April 2002. ISSN 1011-4289. 273 – 277.

3. Kadyrzhanov K.K., Kuterbekov K.A., Lukashenko S.N., Melent'ev M.I., Stromov V.M., Shaitarov V.N. (2000). Assessment of both environment .Borovoye, Kazakhstan. 43-45.

PROBLEM OF RISK MODELING: INFLUENCE OF URANIUM STORAGE ON ENVIRONMENT

A. K. TYNYBEKOV
NGO "International Science Center"
Mcrdist. Dzhal, 94/15, 720038, Bishkek, KYRGYZSTAN

Corresponding author: isc@freenet.kg

ABSTRACT:

Radiological safety of Kyrgyz Republic is caused by natural factors and by activity of the enterprises on extraction and processing of the raw material. It has specific climate and conditions with the highest degree of natural hazards. Increase of level of underground waters, high waters, landslides, high seismicity have resulted the storage destruction, that present threat to the ecology of Central Asia. The resolution of radiological problems of Kyrgyzstan may only be resolved with the complex approach of all neighboring countries and help of foreign experts from donor countries.

Keywords: radiological safety, contamination, uranium, ecology and environment.

INTRODUCTION:

Exploration and operation of uranium deposits on territory of Kyrgyz Republic were made since 1940, and extraction of nuclear fuel was stopped by the end of 1970. During that period of time the uranium deposits were exhausted, and others were conserved. In process of prospecting and search development, industrial and mountain works on uranium objects (mines, cuts etc.) had formed mountain dumps of radioactive breeds. As a result of operation of uranium deposits, enrichment of uranium raw materials have created radioactive dumps and tailings with large quantities of uranium, thorium and other radioactive elements. These problems have imparted to necessity of the decision of a problem of a burial place of radioactive elements and toxic chemical waste products, heavy metals with the minimal risk of environmental contamination. Now in Kyrgyzstan, it has a total of about 25 radioactive tailings and 50 radioactive dumps. Eventually under influence of natural factors there is an aeration, destruction, washout of dumps and tailings. Redistribution of radioactive and toxic chemical substances between radioactive storage and surrounding environment is observed. It results in radioactive and chemical pollution of ground, surface and underground waters, flora and fauna, including places where people work and live.

KADJI-SAI URANIUM TAILINGS:
One of the danger representing radioactive storages is located on the southern coast of Lake Issyk Kul, in a valley called Suhoi-Sai, 2.5 km. to the east from the village Kadji-Sai. The total area of radioactive tailings makes 400000 m^2. The tailing consists of two parts, one half is built up by economic constructions of an electromechanical factory, and another part is location for

gold dump, creating additional load for tailing. In Kadji-Sai uranium oxide was mined not in a traditional way, but from ashes of brown Uranium containing coal from Sogutin deposit. Coal was extracted from local mine and burned producing electric power, and then uranium oxide were taken using acid lightened from ashes. Waste products during manufacture and the industrial equipment were buried underground. Now Kadji-Sai tailing and the protective dam is under influence of natural process and anthropogenous influences. There is a process of destruction of radioactive tailing. Mine Kadji-Sai, is exposed to washout, high waters and midstream that result offset of radioactive materials on the surface, that is one of potential pollutants of southern coast of the Lake Issyk Kul (Pic. 1).

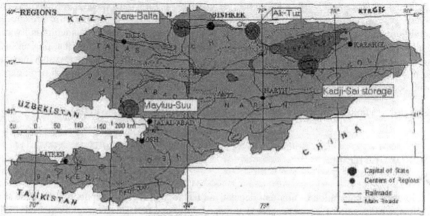

Pic.1 Map of uranium tails

INFLUENCE ON RADIOACTIVE TAILINGS:
Analytical research on susceptibility of radioactive tailings to natural risk were carried out. In our case, we considered the. Ton area, where Kadji-Sai tailing is located. The given area is located in ecologically adverse zone as inhabitants of this region are exposed both to daily and to potential risk.

Concerning natural process, a valley of the river Tone differs by freshet and midstream danger. In it beginning there are 3 lakes (Tuiyk-Tor, Keltor, Korumdy) which are glacial lakes of outburst-danger. The probability of their outburst is increased with global warming of climate, it was received as a result of research, that on Tien-Shan had maximum warming by 0.6 degrees. Per one year it is marked in average to a mountain zone Internal Tien-Shan, minimal on 0.2-degrees to a high-mountainous zone. The greatest warming is during winter-spring months (1.1-0.8) and least in July in a high-mountainous zone 0.1 degree. For a congelation, Tien Shan the negative greatest effect in a similar situation renders rise in temperature in autumn and spring months on 0.5 degree, i.e. causes increase the period of thawing for 30-40 days that conducts to essential acceleration of thawing of glaciers.

Outburst of lakes may cause high waters along a channel of the river the charge up to 2335 m^2/sec. It is dangerous to the population living near the channel and near the rivers. In a zone of destruction after the outburst of these lakes we find area near the shore - a part of the village Kadji-Sai. In Ton region the general area of mudstream danger sites makes 0.16 km^2, with the population more than 200 people. As a result of calculation of a degree of seismic danger of areas Djeti-Ogyz and Ton, more seismically dangerous is Ton region. In the Ton area in a zone of 8-ball seismicity- there are 7 settlements (Kadji-Sai, Bokonbaevo and etc.) and 3170 buildings where 22 thousand people live (Pic. 2).

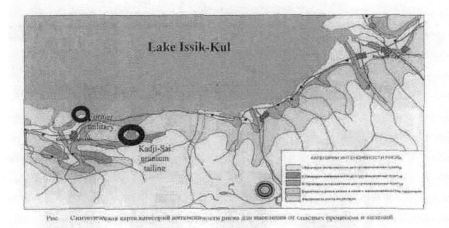

Рис   Сиисмическая карта канчентрации интенсивности риска для населения от спасных процессов и инлений

Pic.2 Risk for the population from natural phenomenas.

Local residents in search for nonferrous metals do carry out excavation in a radioactive burial ground. Artificial destruction of shielding layer of tailings may result in increase of a radiating background. It is possible, that in non-shielded layer of tailing the polluted water and underlying rocks or coastal sites filtered. There is a threat of radiating infection due to a small outflow of uranium waste products from collapsing radioactive tailings. In our case the dangers of the influence of uranium tailings in a zone on an environment is great at occurrence of ecological failure or an accident as the protective dam has low stability. Washout of tailings will result in radioactive and chemical pollution of large territories which may include rivers, arable land, settlements and a coastal zone of the Lake Issyk Kul. The storm rains in 1998 had strongly damaged isolation layer and a dam of the tailing in 1.5 km from the coast of the lake. The radiating background on separate sites has noticeably gone elevated [1].

THE STATE OF THE HEALTH OF THE POPULATION:

The pilot research was carried out to find quality of environmental objects and to study health of the population. The annual reporting of treatment-and-prophylactic establishments on decease of children for 1994-1997 were reviewed and comparing the data on the general children diseases. In Kadji-Sai and Issyk-Kul areas it was established, that the highest parameters on settlement for 1994-1997 were marked by illnesses of breathing, infectious diseases and illnesses of blood which exceeded regional parameters in 1.5-3 times.

In village Kadji-Sai, illnesses are breathing, infectious and parasitic diseases, blood, digestion and endocrine systems. The Comparative analysis by illnesses of respiratory system has shown, that in settlements given pathologies were higher arising as the passage of time than on area as a whole: in 1994 - 2 times, and in 1995 - 3 times, and in 1996 and in 1997 – 3.6 times.

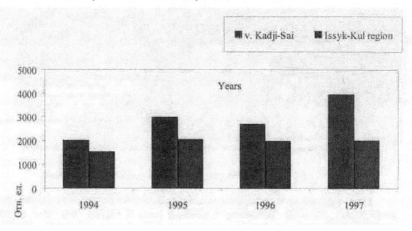

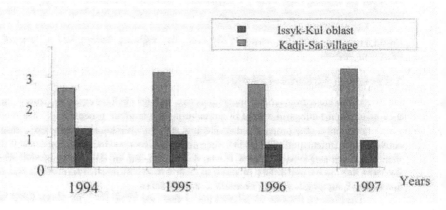

It is revealed, that in dynamics of disease of children in Village Kadji-Sai and Issyk-Kul areas, marked growth was noticed in illness of blood and breathing for 1994-1997 mark growth, and parameters on settlement exceeded regional in 1.3-2 times (Pic. 3).

Thus for the period of 1994-1997, parameter of diseases of children under 14 years of age on 100 thousand children, illness of breathing and blood in village. Kadji-Sai were higher, than on all Issyk-Kul areas (Pic. 4).

RESULT OF RADIOLOGICAL RESEARCH:

For finding - out of a real radiological situation of region and studying of zones with increased radiation, in 1997-2000 research in Issyk-Kul region under grant INTAS and CRDF (USA) were carried out [2]. As a result of this research it was found, that the most part of the investigated territory, the average data of levels of radiation are within the limits of natural radiating background, only in some local sites the increased levels of radiation were fixed. Places with increased indication of a radioactivity were the beach near village Dzhenish, territory of shop #7 near the village Ton and a territory near Kadji-Sai tailing. The increased radiating background in the certain places proves the high contents of radioactive elements, thorium and radium which are products of disintegrating uranium.

During performance of research selective measurements of a level of radiation indoor in various settlements were carried out based on international findings [3]. Results of analyses have shown existence of a significant difference of a level of a radioactivity indoors (Fig. 3).

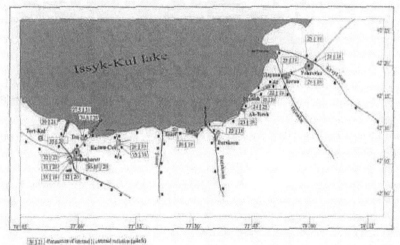

Fig. 3

CONCLUSIONS:

It is necessary to estimate hazard of possible air, water and ground pollution in zone, which is under influence of radioactive storage; to develop technologies and to design the ways of monitoring of ground contaminated by radioactive substances; and to carry out the ecological expertise of the polluted territories.

Acknowledgement: NATO provided full financial support to present this paper at the NATO ARW "Radiation Safety Problems in the Caspian Region" held in Baku, Azerbaijan during September 11-14, 2003.

REFERENCES:

1. Baker SI (1999). Detection of Radon in Rain Water. H.Phys. 77 (5) S-71-76.
2. Hamby DM (1999). A screening assessment of external radiation levels in Kyrghyz Republic, H.Phys. 77 (4). 427-430
3. Sowby D. (2003). Forty years on: How RP evolved internationally. J.Rad.Prot 23. 157-171.

RADIOACTIVE MINERALS AND NUCLEAR FUELS IN TURKEY

DUNDAR RENDA
Mining Consultant, Member of Steering Committee of Turkish Miners Association
Istanbul, Turkey.

Corresponding author: drenda@superonline.com

ABSTRACT:

Nuclear minerals have long been studied in Turkey. These are Uranium and Thorium minerals. Uranium is found in Western and Central Anatolia in sediment and ore deposits and thorium on the other hand is found in complex type deposits in Central Anatolia. Since 1956, Mining Examination and Exploration Institute (MTA) conducted researches, drilling and performed mineral enrichment studies in laboratories and pilot plants for these two nuclear fuel minerals. There is Istanbul-Küçükçekmece Nuclear Research Institute in Turkey, which started its operation in 1955-56. Scientific works are conducted and highly qualified staff is trained in this institute.

Keywords: Nuclear material, Uranium, Thorium, and mining.

INTRODUCTION:

Studies and estimations of relevant authorities and State Planning Organization in Turkey show that shortage that will be experienced in electricity production as a result of increasing consumption of electrical energy, will be possible with nuclear power stations [1].

In the 2nd Five-Year Development Plan (1968-1972), the principle of "Possibilities to take advantage of nuclear energy resources shall be probed and efforts shall be made to launch nuclear power stations" was adopted and to this end in 2nd General Energy Congress that took place in 1968, 400 MW Nuclear Energy Power Plant was discussed and proposed which started its operation in 1977 [1].

Later in 1987, 1992 and 2000 launch of nuclear power plants with 1000 MW, 3000 MW, and 8000 MW were proposed respectively and according to the fixed power it was determined that they would require minimum 210t/a and maximum 714t/a Uranium [2,3]. For the construction of nuclear power plants, which relied on all these studies and information, Mersin-Akkuyu on Mediterranean Coast of South Anatolia was chosen. However, due to environmental problems and other reasons, construction and assembly studies didn't progress. These stations were also expected to reduce the energy unit prices.

M.K. Zaidi and I. Mustafaev (eds.), Radiation Safety Problems in the Caspian Region, 85-88.
© 2004 *Kluwer Academic Publishers. Printed in the Netherlands.*

RADIOACTIVE MINERALS:

The Mining Examination and Exploration Institute (MTA) began research to find reserves of radioactive minerals in Turkey in 1953 and studies progressed until 1999 with some interruptions and funding that was spent within scope of these studies [3-4]. As a result of these works, Uranium reserves with 0.04%-0.1% U_3O_8 contents and Thorium reserves with 0.21% ThO_2 contents were found in Turkey.

Furthermore, in young precipitation formations in Black Sea and Van Lake bases and in Mazıdağı-Mardin phosphate reserves in Southern East Anatolia and in ashes of several lignite in Anatolia, weak U_3O_8 contents of 0.1-200 ppm were detected.

EXPLORATION:

General area studies initiated by MTA in 1953 continued systematically and orderly until the end of 1990. Exploration methods can be listed as scanning broad areas selected according to geological data obtained as a result of field studies through aerial detection, evaluation of abnormalities, exploration drills conducted in promising areas and conducting reserve detection drills according to their results. It should be assumed that Uranium and Thorium reserves found are not the entire reserves of Turkey, and studies conducted in many areas have not been concluded completely.

SUBSTANCE AND ORE DEPOSITS:

URANIUM

9129t Uranium reserves have been found as a result of explorations carried out in Turkey [5]. Reserves in Turkey have generally low U_3O_8 tenor. The most rich reserve is found in Sorgun-Yozgat in Central Asia with U_3O_8 0.1% tenor and a reserve quantity of 3850t. 50% of the reserves consist of one reserve in Salihli-Manisa and 2 reserves in Söke-Aydın, all in Western Anatolia. Reserves in 4 different locations in Aydın and Çanakkale in Western Anatolia and Giresun in Eastern Anatolia and which make up 17% of the entire reserves are small reserves ranging between 250t and 500t and their U_3O_8 contents value between 0.04% and 0.08%. Most of the Uranium reserves in Turkey are sedimentary types. Only one is Gang-type reserve. Gang-type reserves constitute 19% and sedimentary types constitute 81% of the reserves.

THORIUM

As a result of the explorations of Mining Examination and Exploration Institute (MTA), in 1960 a total of 380000t complex Thorium reserve with Barite, Fluorite and Rare Soil Elements in Beylikova-Eskişehir in Central Anatolia was detected [5]. This reserve makes up about 21% of the world Thorium reserve. Average ThO_2 content of found 380000t Thorium reserve is 0.21%. Due to complex structure of the ore deposits, it cannot be used directly as a Thorium deposit. Moreover, explorations to be carried out in Hekimhan-Malatya in Central Anatolia for Thorium will increase Thorium reserves of Turkey.

ORE PRODUCTION, CONCENTRATION AND CONSUMPTION:

The mining activities will get up to speed according to the production of primary energy sources. For the transfer and adaptation of nuclear technology, this will be taken with importance and superiority [6]. Although there are radioactive mineral deposits in Turkey and although some laboratory studies were conducted for their concentration, there are no commercial mine production and concentration gaining ground in Turkey. However, Radioactive Mineral Concentration is imported for the needs of Nuclear Research Center and to be used for medical purposes.

URANIUM

As the continuance of laboratory studies of MTA, a pilot plant was built in 1974 to obtain reliable data for feasibility evaluation and studies were performed for enrichment processes needed for each uranium deposits of Turkey and about 1200 kg Yellow Cake was produced successfully using some of the deposits. Despite the positive results of studies and pilot plants, small amount of U_3O_8 contents of Uranium deposits, small size of reserves makes uranium production and concentration impossible in terms of cost-efficiency. As there are no nuclear fuel reactors in Turkey, not huge amounts of Uranium are wasted. However small amounts of Uranium concentration are imported for medical and technical needs and research of laboratories and Çekmece Research Institute.

THORIUM

Although 380000t Thorium reserves was discovered in 1960 by MTA in complex deposit with Barite, Fluorite, Rare Soil Elements, as a result of the laboratory experiments, technological problems with respect to enriching thorium and separating it from others, could not be overcome. Currently Thorium is a nuclear fuel raw material, which can be used when necessary. It is imported to be used in medicine, chemistry and electronic industry plants.

CONCLUSIONS:

Although radioactive mineral explorations that started in 1953 in Turkey resulted in discovery of 9000t Uranium and 380000t Thorium and although laboratories and institutes have been studying on them for about 50 years to make them available for use, Uranium and Thorium reserves in Turkey, cannot be used because of technological problems and other reasons.

The radioactive raw material, which can be used when necessary, keeps the nuclear fuel character always.

Uranium ore deposits with poor mineral contents and small size of the substance are not feasible for ore production and concentration.

Thorium with rich in mineral contents and on a large scale substance technological problems to enriching and separating ThO_2, could not be overcome

Although plans were proposed, discussed and made, Nuclear Energy Stations, which couldn't be constructed for several reasons, made the production of these radioactive materials impossible.

Turkey closely monitors radiation securities and any likely problems of Nuclear Energy Stations in countries around Black Sea and Hazar Sea Regions and also pays attention to any related technological advancements.

REFERENCES:

1. 2nd Five-Year Development Plan. (1988-1972). Turkish Republic, Department of State Planning Organization, Printed by State Printing House, 1967, Ankara, Turkey.
2. 2nd General Energy Congress of Turkey. (1968). Ankara, Turkey
3. Uranium Exploration Expenses in the World, OECD/NEA and IAEA. (1997). Uranium Resources, Production and Demand, Paris, France
4. Uranium – Thorium, 7th Five-Year Development Plan Report of Special Committee, DPT 2429 - OIK 487. (1996). Ankara, Turkey
5. Uranium - Thorium Exploration in Turkey, MTA Activity Report of the Years 1986–1999. (2000). Ankara, Turkey.
6. 7th Five-year Development Plan (1996–2000). (1995). Turkish Republic, Department of State Planning Organization, Printed by Department of Public Relations and Publications, Ankara, Turkey.

SOLVENT EXTRACTION OF URANIUM FROM WET PROCESS PHOSPHORIC ACIDS

SEREF GIRGIN, AYHAN ALI SIRKECI, NESET ACARKAN
Istanbul Technical University, Mining Faculty, Coal and Mineral Processing Department, 80626, Maslak, Istanbul-Turkey

Corresponding author: sirkecia@itu.edu.tr

ABSTRACT:

In this investigation solvent extraction of uranium from wet process phosphoric acids was realized using industrially available organic extracts 2-Ethylhexyl phosphoric acid (D2EHPA) and Tri-n-octyl phospine oxide (TOPO). The results of initial tests showed that, the effect of different parameters such as uranium oxidation stage, temperature and the molar ratio of D2EHPA/TOPO on the uranium recovery was in good agreement with those of previous investigations. The effect of acid concentration and acid/organic phase ratios suggested that the mechanism of D2EHPA/TOPO synergism was rather complex and it presented different trends.

Keywords: Uranium, extraction, phosphoric acid, wet process and TOPO.

INTRODUCTION:

Worldwide, considerable amount of phosphoric acid is derived from sedimentary phosphate ores for the use of the fertilizer industry. During the process of phosphoric acid production 75 to 90% of uranium in the sedimentary rock reports to the acid [1-3] hence the acid product may contain up to 300 mg/l uranium. Rare metals such as V, Cd and Co and radionuclides like Th and Ra [4] are the other elements associated. There are two major reasons for the extraction of uranium and other elements (thorium, cadmium, cobalt, etc.) from phosphoric acids: to produce uranium as a by-product and to prevent the pollution of soil by the aforementioned radionuclides and rare earth metals through fertilisers [5].

Wet process phosphoric acids are an important uranium resource [6], therefore, several research studies has been conducted on the solvent extraction of uranium from phosphoric acids utilizing organophosphorus acids as the extractant [7-13]. D2EHPA-TOPO couple is the most conventional and practically accepted extractant composition [14,15] and their extraction mechanism have been studied by a number of investigators [11, 16-21]. Although synergistic extraction of species with the mixture of acidic and neutral solvents was explained by the mechanism of addition, substitution, and solvation [22-27], the information given about the parameters affecting the synergism mechanism of D2EHPA-TOPO couple is limited. Therefore, the present research program has been carried out to investigate the effects of such factors as volumetric phase ratio, the concentrations of acid and extractant on the solvent extraction.

M.K. Zaidi and I. Mustafaev (eds.), Radiation Safety Problems in the Caspian Region, 89-96.
© *2004 Kluwer Academic Publishers. Printed in the Netherlands.*

The extractant D2EHPA is a derivative of orthophosphoric acid whose two hydrogen atoms are substituted with two radicals ($R_{1,2} = C_8H_{17}$). It was reported that U (VI) was extracted by acidic extractants in a dimeric form of $UO_2 (HA)_{2A2}$ [28, 29] where A represents the alkyl ($-OC_8H_{17}$) and phosphoryl groups (P = O) of a typical organophosphorus acidic extractant and HA is the organophosphorus acidic extractant itself. Furthermore, whether the metal-extractant complex formed will be a monomer or polymer depends on metal loading level of the extractant.

TOPO is a neutral donor, and is synthesized by substituting 3 hydroxyls in the chemical structure of orthophosphoric acid by 3 organic radicals. Uranium is extracted by TOPO through the coordination with the oxygen of the phosphoryl group (P = O) in the structure. TOPO does not release any hydrogen ions as a result of dissociation; therefore, extraction is not affected by the acidity of the solution as opposed to the case of D2EHPA. The chemical structures of D2EHPA and TOPO are given as follows:

$$
\begin{array}{ccc}
O=P & \begin{array}{l} \diagup OH \\ \!\!\!\!\!\!\text{—— OR} \\ \diagdown OR \end{array} &
O=P & \begin{array}{l} \diagup R \\ \!\!\!\!\!\!\text{—— R} \\ \diagdown R \end{array} & R: C_8H_{17}
\end{array}
$$

$$\text{D2EHPA} \qquad\qquad \text{TOPO}$$

EXPERIMENTAL:

Materials:

Experimental tests were carried out using a wet process phosphoric acid sample from Morocco that contained 53% P_2O_5 and 205mg/l uranium. The chemical composition of the test sample is presented in Table 1.

Table 1. Chemical composition of the test sample.

U	205 mg/l	Ni	60 mg/l
V	282 mg/l	Fe_{total}	3.38 g/l
Th	4 mg/l	Fe^{+2}	1.49 g/l
Ti	114 mg/l	P_2O_5	53.17%
Cd	30 mg/l	H_3PO_4	73.42%

Analytical grade TOPO and D2EHPA were used as organic solvents whereas technical grade kerosene was the diluents. Extraction was realized in a 1.5-liter capacity double jacket extraction vessel where the liquid phase was mechanically agitated. Temperature was maintained within ± 0.1^0 C by means of a water bath and electronic control unit.

Method:

In this investigation, the effect of acid concentration, organic concentration and the volumetric ratios of acidic and organic phases on the extraction mechanism of uranium from the wet process phosphoric acid were studied. Tests were conducted employing 1.5, 2.4, 3.3, 4.3, 6.6 M H_3PO_4 concentrations with corresponding uranium concentrations of 26, 39, 56, 79, 97 mg/l, respectively. Preliminary test results suggested that a 3-minutes agitation time at 400 rpm was adequate. All tests were carried out at 35^0C.

The results were evaluated in terms of uranium extraction coefficient (E) and uranium recovery (R,%). The following equation depicts the extraction coefficient E:

$$E = \frac{\text{Uranium concentration in organic phase}}{\text{Uranium concentration in aqueous phase}}$$

The phosphoric acid sample was clarified with activated bentonite in order to prevent crud formation during the extraction process. The oxidation level of uranium was controlled using sodium chlorite ($NaClO_3$) dissolved in distilled water. Uranium in the acidic phase was assayed by a UV spectrophotometric method using 2-(5-brom-2-pyridilazo)-5-(diethyl amino)-phenol as the indicator. Other elements in the phosphoric acid sample were analysed using the ICP method.

RESULTS AND DISCUSSION:

TESTS WITH INDIVIDUAL SOLUTIONS OF D2EHPA AND TOPO

Tests carried out with the individual solutions of D2EHPA and TOPO at 2/1 phase ratio indicate that extraction abilities of both extractants get poorer as the acid concentration increases (Table 2).

Table 2. The effect of acid and organic concentrations on the uranium extraction (aqueous/organic = 2/1).

Organic Concentration, M	H_3PO_4 Concentration, M				
	1.5	2.4	3.3	4.3	6.6
D2EHPA	R, %	R, %	R, %	R, %	R, %
0.7	92.0	78.4	55.7	32.4	21.3
1.0	96.2	87.3	66.6	35.0	23.1
1.5	98.4	92.6	73.5	60.6	30.2
TOPO					
0.38	25.9	17.7	12.2	8.8	4.4

Test Carried Out With the Mixture of D2EHPA and TOPO

Systematic tests were conducted with different concentrations of acid and varying concentrations of extractant (D2EHPA+TOPO) at different phase ratios (aqueous/organic). At all extractant concentrations the molar ratio of D2EHPA to TOPO was 4.

Effect of Acid and Organic Concentration at Phase Ratio of 1/1

At 1/1 phase ratio, uranium recovery increases in parallel to the organic concentration and eventually 98% extraction is achieved at 2.4M H_3PO_4 concentration (Table 3). As the acid concentration increases uranium recovery drops to a minimum of 68%.

Table 3. The effect of D2EHPA and TOPO concentrations
on the uranium extraction (phase ratio 1/1).

Organic Concentration, M D2EHPA+TOPO	H_3PO_4 Concentration, M					
	2.4		4.3		6.6	
	E	R, %	E	R, %	E	R, %
0.5+0.125	4.6	85	---	---	---	---
0.7+0.18	6.2	86	2.6	72	2.1	68
1.0+0.25	32.0	97	3.4	77	2.7	73
1.5+0.38	85.0	98	5.1	84	4.7	82

Effect of Acid and Organic Concentration at Phase Ratio of 2/1

Preliminary tests conducted using 1M D2EHPA only at a 6.6M acid concentration and 2/1 phase ratio showed that the uranium recovery (R) was 23.1%. On the other hand, in the existence of 0.25M TOPO uranium recovery increases up to 62% corresponding to 168% improvement due to synergism (Table 4).

Table 4. The effect of organic and acid concentration on the uranium extraction (phase ratio 2/1).

Organic Concentration, M D2EHPA+TOPO	H_3PO_4 Concentration, M									
	1.5		2.4		3.3		4.3		6.6	
	E	R,%	E	R,%	E	R,%	E	R,%	E	R,%
0.5+0.125	---	---	4.9	71	---	---	---	---	---	---
0.7+0.18	14.1	88	10.3	82	6.0	75	3.3	62	2.6	57
1.0+0.25	74.0	97	55.3	95	13.9	87	4.9	71	3.2	62
1.5+0.38	127.0	98	85.0	97	16.6	89	6.8	77	5.2	72

As the acid concentration decreases, uranium recovery increases up to 98% especially below 3.3M H_3PO_4 concentration.

At high acid concentration (6.6M) 21.3% and 4.4% uranium recovery is obtained when 0.7M D2EHPA only and 0.38M TOPO only is used respectively (Table 2. However, when 0.7M D2EHPA and 0.18M TOPO are used together the uranium recovery goes up to 57% (Table 4) that is comparable to the results obtained with individual extractants. Since there is no H^+ liberation at 4.3 to 6.6M H_3PO_4 concentration range for 0.7M D2EHPA + 0.18M TOPO, it could be deduced that TOPO strongly promotes the extraction ability of D2EHPA without causing any H^+ ion liberation.

TESTS CONDUCTED WITH A PHASE RATIO OF 4/1

The test results show that as acid and extractant concentrations are increased, uranium recovery goes up to 90% (Table 5).

Table 5. The effect of organic and acid concentration on the uranium extraction (phase ratio 4/1).

Organic Concentration, M D2EHPA+TOPO	H_3PO_4 Concentration, M					
	2.4		4.3		6.6	
	E	R, %	E	R, %	E	R, %
0.5+0.125	6.5	62	---	---	---	---
0.7+0.18	13.2	75	3.5	47	3.1	43
1.0+0.25	20.6	82	4.8	54	4.4	53
1.5+0.38	39.0	90	7.3	65	5.2	57

Tests Conducted With a Phase Ratio of 8/1

At 8/1 phase ratio, where metal loading in the organic phase is relatively maximum, the effect of extractant and acid concentration on the uranium recovery presents a different trend than those observed for other phase ratios (Table 6). Therefore the reaction mechanism at 8/1 phase ratios differs from the lower phase ratios investigated in this study.

Table 6. The effect of organic and acid concentration on uranium extraction (phase ratio 8/1).

Organic Concentration, M D2EHPA+TOPO	H_3PO_4 Concentration, M					
	2.4		4.3		6.6	
	E	R, %	E	R, %	E	R, %
0.5+0.125	3.9	33	---	---	---	---
0.7+0.18	6.3	44	4.9	38	3.6	31
1.0+0.25	14.9	65	3.5	30	3.6	31
1.5+0.38	14.9	65	5.7	42	7.8	49

DISCUSSION:

Figures 1 and 2 show a summary of test results conducted with different acid and organic concentrations at 2/1 phase ratio. It seen from Figure 1 those considerable differences exist in extraction coefficient between different extractant concentrations. While E is around 14 at lowest extractant and 1.5 M acid concentrations it becomes 74 and 127 at increasing extractant concentrations. However, the figures for uranium recoveries are rather different. At the same

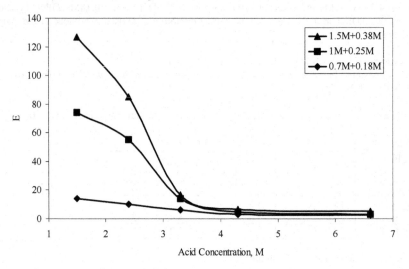

Figure 1. Extraction coefficient against acid concentration at 2/1 phase ratio.

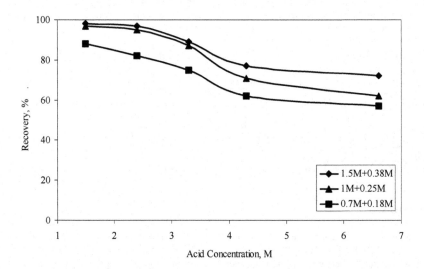

Figure 2. Uranium recovery against acid concentration at 2/1 phase ratio.

conditions recoveries are 88, 97 and 98% respectively. When extraction coefficient is taken into consideration 1.5 + 0.38M extractant concentration seems to be ideal for the efficient removal of uranium yet marginal increase in the recovery is minimal hence 1 + 0.25M extractant concentration would be more appropriate from the economical point of view. Acid concentrations up to 3.3 M may be acceptable above that recovery drops below 80%, however, the reduction in extraction coefficient is immense. On the other hand, while extraction coefficients only are not helpful for decision making regarding optimum conditions for extraction process, they provide invaluable information about the mechanism of the extraction. Those deductions with respect to extraction coefficients and extraction mechanisms were discussed in detail elsewhere [30].

CONCLUSIONS:

Reaction mechanisms of uranium extraction from wet process phosphoric acid sample using D2EHPA+TOPO synergistic couple depend on the acid and extractant concentrations.

Uranium extraction mechanism for D2EHPA differs when TOPO exists in the organic phase.

At relatively low initial acid and high D2EHPA+TOPO extractant concentrations, extraction mechanism proceeds towards D2EHPA dominant cation exchange mechanism.

As the acid concentration is increased, TOPO becomes more and more effective on the synergism thus improving the uranium recovery.

Acid concentration has a significant role on the extraction of U (VI) from H_3PO_4 such that at 4/1 and lower phase ratios the critical acid concentration is 3.3M above that distribution coefficient considerably drops and sufficient extraction cannot be achieved.

Acknowledgement: NATO provided full financial support to present this paper at the NATO Advanced Research Workshop, Radiation Safety Problems in the Caspian Region, held in Baku, Azerbaijan during September 11-14, 2003.

REFERENCES:

1. Ring, R.J. (1977) Jan. Atomic Energy in Australia. 12.
2. McGinley, F.E. (1979) Oct. IAEA-SM-239/15, Buenos Aires.
3. Stevens, (1980). D.N. Uranium Resources/Technology Seminar III, March 10.
4. Scholten, L.C., Timmermans, C.W.M. (1996). Fertilizer Research, 43, 103-107.
5. Ionization Radiation: Sources and Biological Impact. UNO Scientific Committee on Nuclear Radiation Hazard. Report to the UNO General Assembly (1982).1. 257, 293, 413.
6. OECD 1999, Environmental Activities in Uranium Mining and Milling, Report by OECD-Nuclear Energy Agency and IAEA. (1999). 29.
7. Fitouss, R., Musikas, C (1980). Sep. Sci. Tech. 15. 845.

8. Mouysset, G., Moliner, J., Lenzi, M. (1983). Hydrometallurgy. 11. 165.
9. Noirot, P.A., Wosniak, M. (1985). Hydrometallurgy. 13. 229.
10. Daoud, J.A., Zeid, M.M., Aly, H.F. (1997). Solvent Extraction and Ion Exchange. 15 (2). 203-217.
11. El-Reefy, A.S., Awwad, S.N., Aly, H.F. (1997). J.Chem.Tech. and Biotech. 69 (2). 271-275.
12. Krea, M., Khalaf, H. (2000). Hydrometallurgy, 58, 215-225.
13. Singh, H., Vijayalakshmi, R., Mishra,S.L., Gupta C. K. (2001). Hydrometallurgy. 59. 69-76.
14. Mcready, W.L., Wethington, J.A., Hurst, F.J. (1981). Nucl Tech. 53. 344.
15. Hurst, F.J. (1989). The Recovery of Uranium from Phosphoric Acid. Int. Atomic Energy Authority, Vienna TEC DOC, 533.
16. Hurst, F.J., Crouse, D.J., Brown, K.B. (1969). Report ORNL-TM-2522.
17. Hurst, F.J., Crouse, D.J., Brown, K. (1972). B. Ind. Eng. Chem. Process. Des. Dev. 11. 122.
18. Bunus, F.T., Domocos, V.C., Dimitrescu, P.J. (1978). Inorg. Nucl. Chem. 40. 117.
19. Bunus, F.T., Miu, I., Dimitrescu, P.J. (1994). Hydrometallurgy. 35. 375.
20. Dahdouh, A., Shlewit, H., Khorfan, S., Koudsi, Y. (1997). Journal of Radioanalytical and Nucl. Chem. 221 (1-2).183-187.
21. Rafajfeh, K.M., Al-Matar, A.Kh. (2000). Hydrometallurgy. 56. 309-322.
22. Baes C.F. (1963). Jr, Nucl. Sci. Eng. 16. 405.
23. Vdovenko, V.M., Krivokhatskii, A.S. (1960). Zh. Neorg. Khim. 5. 494.
24. Dyrssen, D., Kuca, L. (1960). Acta Chem. Scand. 14. 1945.
25. Kennedy, J. (1958). AERE Report C/M Harwell.
26. Irving, H.M.N.H., Eddington, D.N. (1965). J.Inorg. Nucl. Chem. 27. 1359.
27. Healy, T.V. (1961). J. Inorg. Nucl. Chem.19. 314.
28. Kosolpoff, G.M., Powell, J.S. (1950). J. Cem. Soc. 3535.
29. Baes, C.F., Zinger Jr, A., Coleman, C.F. (1958). J.Phys. Chem. 62. 129.
30. Girgin, S., Acarkan, N, Sirkeci, A.A. (2002). J.Radioanalytical and Nuclear Chemistry. 251 (2). 263-271.

RADIOACTIVITY OF LAKES IN THE URBANIZED TERRITORIES

V.A. MAMMADOV
Geology Institute, Azerbaijan National Academy of Sciences
H. Javid av., 29A, Baku-370143, Azerbaijan.

Corresponding author: radiometry@gia.ab.az

ABSTRACT :

The role of Anthropogenic factors during the accumulation and migration of radionuclides in saline lakes contaminated by oil-products in the Absheron peninsula were studied. The more high indicators of radioactivity manifest in bottom sediments of the lakes. In the bottom deposits of Gala-2 Lake it was discovered the abnormally high contents of radium. In water, the radionuclides contents   is low as likely they deposit by time. There was also determined that natural diversity of these factors is a strong modification of the accumulation and migration of radionuclides in the lakes.

Keywords: morphometric, biotopes, ecolimnologic and typological character.

INTRODUCTION:

While evaluating recent state of environment and lakes in particular which are situated in the urbanized territories their radioactivity is of a great importance. In Azerbaijan the Absheron peninsula is among urbanized territories exposed to active Anthropogenic pressing. Since the 19th century lakes in the Absheron peninsula have been suffering stage-by-stage and various Anthropogenic impacts. This was facilitated by intensive development of oil-producing and oil-chemical industries, by the growth of populated areas and agro-industrial complexes and also by many other events. Absence of refining complexes in factories as well as of central sewage system on many populated areas in the enormous draining of non-purified waste into lakes. On the one hand, the growth of amount of waters lead to the change of morphometric and physical-chemical parameters of lakes and also number and species of organisms existing therein and their biotopes. On the other hand there appeared many artificial draining lake-like reservoirs with heavily polluted waters.

METHODS:

To study radioactivity of the lakes in the Absheron peninsula in 1998 and in 2000 we took specimens of water and bottom sediments from 20 typical lakes. Selection site of bottom specimens is: coast-center of the lake, from the surface of the floor as deep as 0,5 m. There was determined amount of thorium, radium, potassium, radon-220, radon-222 in Bk/l and in the bottom sediments there was determined amount of uranium (in compliance with radium),

M.K. Zaidi and I. Mustafaev (eds.), Radiation Safety Problems in the Caspian Region, 97-102.
© 2004 *Kluwer Academic Publishers. Printed in the Netherlands.*

thorium, potassium in Bk/kg. These analyses were conducted at the Geology Institute of Azerbaijan, National Academy of Sciences on the gamma-spectrometric installation ASRM-2 and also developed the methods. The location of lakes understudy are shown in the figure, posted below.

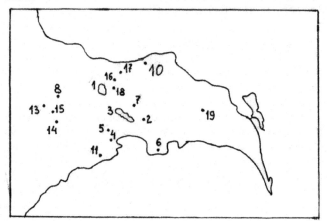

Figure. Location of the studied lakes on the Absheron peninsula.

RESULTS AND DISCUSSION:

Process of Anthropogenic contamination of lakes in the Absheron peninsula (by radionuclides inclusive) started in the 19th century but in the 20th century it became more intensive and irreversible. Ecolimnologic studies in different countries of the last decade demonstrate that even partial contamination results in a serious genetic reconstruction of the whole system of processes of mass- and energy exchange in a lake-like reservoir. With a course of time it changes its whole typological character. The lakes became more vulnerable than ecosystems of seas, oceans and rivers [1].

Given below are some examples illustrating the growth of radioactivity of lakes. In the beginning of the 19th century the daily discharge of the surrounding lakes in oil fields into the lake of Beyuk Shor was 2-3 th.m^3 of contaminated waters. By the 70s-80s their amount grew up to 30 th.m^3 and neighboring territories were flooded.

The lakes Gala, Gala-2, Zabrat, Sianshor, Gyrmyzy, Puta, Ramana, Byulbyulya etc. are flooded by oil and sewage waters. In soils of water-reservoirs, on banks and coasts, in water and in the floor of these lakes there was accumulated a huge amount of oil-products and other contaminants. The chemical analysis of bottom deposits revealed that there are oil-products in many studied lakes. The lakes Beyuk-Shor – 300-400 gr/kg, Zabrat – 110-260 gr/kg, Gala – 170-250 gr/kg, Gyrmyzy – 35-270 gr/kg, Byulbyul – 30-190 gr/kg, Pirshagi – 90-150 gr/kg, Masazyr – 10-80 gr/kg, Duzlu – 25-70 gr/kg, Khodjagasan – 30-90 gr/kg etc. have a large amount of oil-products. Less amount of oil-products (about 1-30 gr/kg) is observed in the bottom deposits of lakes of the west part of Absheron. In some areas in drainage-waters reservoirs oil products penetrate intro the soils as deep as 1,5-2,0 m. Amount of oil-products in the water often exceeds PC tens and hundreds times. It should be mentioned that in all stratal waters in oil fields there exists uranium.

In oil-saturated sands and sandstones in comparison with its "empty" analogue it is observed the abnormal content of uranium, radium, vanadium, chrome, cobalt, barium, strontium, beryllium. In oil saturated clays in comparison with usual clays one can observed: a) increased concentration of uranium, radium, titanium, vanadium, chrome, manganese, cobalt, copper, beryllium, strontium, tin, plumbum; b) low concentration of potassium, thorium, nickel [2]. The lakes contain the same radioactive elements as rivers and seas. Main radioactive elements in natural waters are uranium, radium, radon and potassium. The first data of radioactivity situation of urbanized territories of Azerbaijan was carried out in 1999 on the example of Absheron peninsula lakes [3]. Results of investigations [4] demonstrate that "normal radioactive background in the Absheron peninsula varies 4-10 mc Rt/hour. In outcrops of Oligocene-Miocene layers (Maikopian suite) it grows to 20-25 mc Rt/hour. In fields of mud breccias of volcanoes Zigilpiri, Abich, Keireki, Beyukdag it grows up to 12-15 mc Rt/hour. Along the north coast of the Absheron peninsula and in the Shakhov spit radioactivity drops to minimum of 3 mc Rt/hour in zones of beaches. Average background level in the region varies about 6 mc Rt/hour". These authors explain low level of natural radioactivity in the Absheron peninsula as compared with other regions in Azerbaijan by geologic structure. The Absheron peninsula is a young geologic province. Most of its territory is composed of recent sedimentary weakly radioactive deposits represented by clays, sandstones, sands and limestone.

Results of the analyses to verify amount of radionuclides in the water of the investigated lakes demonstrate their absence in the water. Only one analysis revealed a small amount of radium, in the water of the lake Khodjagasan (see Table). Radioactivity of some lake waters in the Absheron peninsula is determined by potassium. Amount of potassium varies in a wide range – from 0 to 1,4 g/l. In most of the lakes amount of potassium is not more than 0,2-9,3 g/l. 0,56 g/l of potassium in the lake Pirshagi and 1,2-1,4 g/l of potassium in the lake Gyrmyzy are abnormal and demonstrate high concentration of salts in the waters of these lakes. Probably potassium is delivered to the lakes with coastal drainage. However one can suppose that the lake Gyrmyzy is a relict of marine waters and it is a basin of a lagoon origin [5]. In the oceanic water amount of potassium varies about 0,39 g/l, i.e. it is rather high to create anomaly in the waters of the relict lakes during the evaporation. In the lakes like Beyuk Shor, Masazyr, Duzly etc. amount of potassium is somewhat lower than in the oceanic water but many times it exceeds its amount in the river water.

Bottom sediments of lakes turned out to be more radioactive than the water series. Due to a high absorptivity they (mainly oozy deposits with rich organics) accumulate significant amount of radionuclides, which are delivered into the lake. Besides Mirzaladi, Sianshor, Dashgil-lesser and Duzlu lakes uranium deposits exist in all other lakes but its amount does not exceed the PC.

Potassium exists in all types of the bottom sediments (except for abnormal lake Gala-2). Notwithstanding that amount of potassium is lower there than the clark of sedimentary deposits this demonstrates its uninterrupted delivery from the environment.

Table: Amount of radionuclides in bottom sediments in the lakes of Absheron

№	Lake	Selection site and peculiarities of the deposits	Activity	Amount of radionuclides						Contribution of radionuclides in general gamma-radiation, %		
				Bk/kg			%					
				U (Ra)	Th	K	$U \cdot 10^{\text{□4}}$	$Th \cdot 10^{\text{□4}}$	K	U	Th	K
1	2	3	4	5	6	7	8	9	10	11	12	13
1	Masazyr	Centre, oozy grey sand	42,8	12,1	7,1	272,6	0,98	1,75	0,92	47	10	43
2	Byulbyulya	Coast, ooze, loams with oil	30,8	13,2	0	224,1	1,07	0	0,76	49	0	51
3	Beyuk-Shor	Coast, ooze, sand with fuel oil	16,1	26	0	171,9	0,21	0	0,58	19	0	81
4	Ganlygel	Centre, dark grey-green ooze	70,8	47,7	5,2	207,7	3,86	1,27	0,70	71	9	20
5	Khodjagasan	Shore, ooze, with a smell of hydrogen sulphide	33,7	27,8	0	76,2	2,24	0	0,25	85	0	15
6	Duzlu	Centre, black-grey ooze	11,9	0	0	151,6	0	0	0,51	0	0	100
7	Zabrat	Centre, clayey ooze	89,6	52,9	20,8	120,5	4,28	5,29	0,41	62	29	9
8	Dashgil	Coast, oozy clay with bituminiferous deposits, grey with a smell of oil and sulphur	81,2	39,1	12,5	329,0	3,17	3,05	1,11	52	20	28
9	Dashgil, lesser	Coast, mixture of ooze and fuel oil, black-brown with a sharp smell of oil	54,6	0	33,2	141,6	0	8,15	0,48	0	82	12
10	Pirshagi	Centre, grey-	53,3	21,6	7,3	282,5	1,75	1,79	0,95	45	18	37

		green ooze										
11	Gyrmyzy, main	Coast, ooze, clay, sand, fragments of coquina with oil-fuel oil mixture	43,7	29,8	0	241,1	2,01	0	0,81	62	0	38
12	Gyrmyzy, main	Centre, mixture of ooze with fine-grained sand, black-grey	30,4	15,4	0	190,5	1,25	0	0,64	56	0	44
13	Shorchala, west	Centre, brown-yellow ooze	56,3	21,6	0	442,9	1,75	0	1,50	44	0	56
14	Meiliguluchala	Centre, black ooze	43,6	17,8	0	327,9	1,45	0	1,08	47	0	53
15	Eirichalashor	Centre, grey-yellow ooze	40,1	9,7	0	387,4	0,79	0	1,31	29	0	71
16	Sianshor	Coast, black ooze with fuel-oil	17,5	0	0	223,4	0	0	0,75	0	0	100
17	Fatmai	Centre, grey-black ooze	38,1	20,9	0	220,1	1,69	0	0,74	60	0	40
18	Mirzaladi	Centre, black-blue ooze with a smell of oil	12,5	0	5,7	64,0	0	1,4	0,22	0	63	37
19	Gala-2	Coast, black ooze with fuel-oil	564,6	418,0	111,9	0	33,8	27,47	0	75	25	0

Typical range o specific activity of potassium is 110-740 Bk/kg (Moiseyev, Ivanov, 1990). In the bottom deposits of the lakes in Absheron its amount is 64-443 Bk/kg.

Thorium is rare in the bottom deposits of the lakes. Uranium does not exceed clark for sedimentary rocks as well and it varies 0-52 Bk/kg. However in lakes Zabrat, Ganlygel, Dashgil there still occurs accumulation of uranium and lake Gala-2 is heavily contaminated by uranium and thorium. In lakes with increased amount of uranium contribution of potassium becomes acutely lower. There is also a direct relation between activity of bottom deposits and uranium content. Contribution of thorium haven't influence on the integral radioactivity.

Lake Duzlu is the least radioactive (it is situated near the Caspian Sea shore) – 0,3-0,4 mc Rt/hour and it is higher in the centre (11,2) than near the shore (9,7). Change of radioactivity along line from the lake bed to the native coast is not always even and in one direction.

Relatively increased radioactivity in some biotopes of lakes demonstrate their local contamination. In the lakes radionuclides are accumulated in organisms and while moving along the alimentary line lead to their oppression.

The above given data on radioecologic investigations of the lakes in the Absheron may promote better understanding of the Anthropogene impact on the environment while taking measures to protect and to utilize lacustrinal systems.

CONCLUSION:

Radioecological studies of lakes can promote to more detailed study of degree of human being influence upon nature during the solution of practical problem of defense and use of lacustrine ecosystems. Contamination of lakes of Absheron peninsula during last 150 years had led to that the anthropogenic factor in the lake regime began to be the leading one in comparison with natural. In bottom deposits of very contaminated with oil-products lakes the radioactivity is higher, than in water medium. There is not clear relation between concentration of radioactive elements and mineralization of lake waters.

REFERENCES:

1. M.Ya. Chebotina, N.V. Kulikov. (1998). Ecologic aspects of the study of migration of radionuclides in continental basins. Ecology. 4. 282-290 (in Russian).
2. Ch.S. Aliyev (1994). Ph.D. thesis "The radioactive fields of the depression zones of Azerbaijan". p. 337 (in Russian).
3. V.A. Mammadov. (1999). About radioactivity of lakes in Absheron. Izv. NAN Azerbaidzhana, Earth Sciences, № 1, p. 123-128 (in Russian).
4. Ch.S. Aliyev, T.A. Zolotovitskaya, M.V. Podoprigorenko (1996). Radionuclide contamination of the environment when developing oil fields. ANH. 7. 46-50 (in Russian).
5. Ch.S. Aliyev, T.A. Zolotovitskaya, V.A. Mammadov (2002). Processes of contamination of lakes in the low land–piedmont areas in Azerbaijan; radioactivity of waters and bottom deposits. Izv. NAN Azerbaidzhana, Earth Sciences, № 1, p. 81-89 (in Russian).

NATURAL RADIONUCLIDES IN SOIL-PLANTS IN SHEKI-ZAKATALA ZONE OF AZERBAIJAN

A.A. GARIBOV, I.A. ABBASOVA. M.A. ABDULLAYEV, CH.S. ALIYEV
Institute of Radiation Problems, Azerbaijan National Academy of Sciences
H. Javid Avenue 31 a, AZ1143, Baku, AZERBAIJAN.

Corresponding author: hokman@rambler.ru

ABSTRACT:

Tobacco is one of the most usable plant and its goods are utilized in the production of medicines. Lately, several natural radioactive elements contained in tobacco caused the greatest anxiety. The research using a gamma-spectrometric and radiometric link the regularity of the distribution of the natural radionuclides (U-238, Ra-226, Th-232, K-40) have been described due to the scheme: about soil ▭ tobacco ▭ stalk of tobacco ▭ leaves, in the northern region of Azerbaijan. The regularity of free radicals influence on the density of natural radioactive isotopes has been discovered in tobacco.

Keywords: Tobacco, medicine, radionuclides, uranium, thorium and radium.

INTRODUCTION:

Natural soil radioactivity was for the first time investigated in former Soviet Union in 1936. Further these investigations of natural radionuclides spread in soil was investigated at the Institute of Geography [1]. Natural radionuclides spread in the grounds of different regions of Azerbaijan, their accumulation in various crops and the influence of mineral fertilizers accumulation in crops had been widely investigated [2-4,5]. The research dealing with decreasing their harmful influence are being done.

The knowledge about the amount of the natural radionuclides in the tobacco gives the opportunity that they should be concerned to the low radioactive factors, which characterize the problem of studying of noxious features for human beings. That is why depending on the regions where tobacco is grown revealing the regularity of conferment of the natural radionuclides by tobacco plant has the greatest importance. Radionuclides have been researched by the methods of gamma-spectrometric and radiometric. The results show that the accumulation of K-40 isotope in tobacco is the most effective, the specific activity decreases from root to leaf, but in dry leaves it is high. The regularity of free radicals influence on the density of natural radioactive isotopes has been discovered in tobacco. The results show that the accumulation of K-40 isotope in tobacco is the most effective, the specific activity decreases from root to leaf, but in dry leaves it is high.

M.K. Zaidi and I. Mustafaev (eds.), Radiation Safety Problems in the Caspian Region, 103-106.
© 2004 *Kluwer Academic Publishers. Printed in the Netherlands.*

The main objective of the work was studying of natural radionuclides spread in soil and their accumulation in different plants of Sheki-Zakatala region of Azerbaijan.

MATERIAL AND METHOD:

Natural radionuclides migration in Sheki-Zakatala region was investigated by indication soil cuts via comparative-geographical method. Indication of soil cuts were done till soil-formative layer. The radionuclides quantity in soil samples was determined by physical method by means of SARI-2 gamma-spectrometer. The samples were placed in the volume of 200g in special cells completely absorbing gamma rays of natural background. The activity of every kind of radionuclides was measured according to gamma spectrum of the samples.

RESULTS AND DISCUSSIONS:

The results according to natural radionuclides accumulation quantity in soil of Sheki-Zakatala region have been shown in Table-1.

Table 1. The quantity of natural radionuclides in Sheki-Zakatala region's soil:

Regions	Cut depth, cm	Specific activity	^{238}U	^{232}Th	^{40}K
			Bq/kg		
Sheki	0-20	47,80	5,08	11,27	356,45
	20-40	59,14	31,05	-	358,25
	40-60	52,63	25,42	-	347,08
Average		53,19	20,52	11,27	353,93
Sheki	0-20	34,51	3,89	2,90	342,08
	20-40	44,76	2,32	13,92	308,70
	40-60	29,96	-	8,92	232,47
Average		36,41	3,11	8,58	294,42
Zakatala	0-20	56,32	20,82	0,15	450,17
	20-40	67,79	19,51	7,51	490,24
	40-60	76,39	-	25,82	542,97
Average		66,83	20,17	11,16	494,46
Zakatala	0-20	57,99	36,55	1,57	247,10
	20-40	54,85	16,24	4,76	412,94
	40-60	51,56	15,79	9,10	304,19
Average		54,80	22,86	5,14	321,41

As it is seen from the table, the specific activity of soil changes within the following range 53,19-36,41 Bq/kg in Sheki region, 76,39-54,80 Bq/kg in Zakatala region. Zakatala region's soil differs in its high activity from that one of Sheki region. One can come across with the highest quantity of Uranium-238 radionuclides in Zakatala region's soil. The soil of the given region is characterized by the highest quantity of thorium 232 and potassium-40.

The results according to the quantity of natural radionuclides in plants of Sheki-Zakatala region are represented in Tables 2-3. The quantity of chemical or radioactive elements in plants is always measured by their accumulation factors in biogeochemistry. Accumulation factor equals to the quantity of element or radionuclides in plants is divided by quantity in soil. Accumulation coefficient of radionuclides in the plants at soil-climate condition depends upon the type or sort of plant and change from 0.01 to 15.0 or more [1,4-5].

Table 2. Accumulation of natural radionuclides in plants of Zakatala region and their accumulation coefficients (AC)

Plant	Sort	Parts of Plant	Specific activity	^{238}U		^{232}Th		^{40}K	
			Bq/kg	Bq/kg	AC, $n*10^{-2}$	Bq/kg	AC, $n*10^{-2}$	Bq/kg	AC, $n*10^{-2}$
Tobacco	Trapezond-15	Root	13.79	10.45	52	.	-	42.64	9
		Stalk	2.87	-	-	-	-	3669	7
		Wet leaf	5.00	4.29	21	-	-	9.08	2
	Samsun-155	Stalk	4.35	-	-	-	-	55.58	11
		Wet leaf	10.98	9.94	49	-	-	13.28	3
	Zakatala-67	Wet leaf	1.42	-	-	-	-	18.13	4
		Dry leaf	51.21	-	-	-	-	653.3	132
	Zakatala	Stalk	3.29	-	-	1.15	10	22.7	5
	Large leave	Wet leaf	3.48	3.48	17	-	-	-	-
Meadow plants	Stalk		3.26	1.37	7	-	-	23.73	7
Tobacco	Zakatala-67	Root	37.7	35.11	154	-	-	33.05	10
		Stalk	6.89	0.40	2	-	-	82.68	26
		Wet leaf	2.94	-	-	-	-	37.51	12
	Samsun-155	Root	7.24	3.34	4	-	-	49.81	15
		Stalk	4.96	-	-	-	-	63.32	20
		Wet leaf	4.92	-	-	-	-	62.76	20
	Zakatala Large leave	Root	4.66	0.99	4	-	-	46.91	15
		Stalk	20.72	18.54	81	-	-	27.38	9
		leaf	21.69	16.05	70	-	-	71.81	22
	Trapezond-15	Stalk	17.10	14.95	65	-	-	27.34	9
		leaf	10.81	5.86	26	-	-	63.16	20
Meadow plants	Root		59.04	49.07	215	-	-	127.1	40
	leaf		4.98	-	-	3.81	74	-	-

Table 3. Accumulation of natural radionuclides in plants of Sheki region and their coefficients (AC)

Plant	Sort	Parts of Plant	Spec ific activ ity	^{238}U			^{232}Th		^{40}K	
			Bq/kg	Bq/kg	AC, n*10^{-2}	Bq/kg	AC, n*10^{-2}	Bq/kg	AC, n*10^{-2}
Tobac co	Zakatala-67	Root	28.4	14.2	69	-	-	180.9	51
		Stalk	12.1	-	-	-	-	153.46	43
		Wet leaf	10.6	0.8	4	-	-	124.33	35
	Zakatala-Dubek	Root	12.8	-	-	-	-	163.67	46
		Stalk	77.5	77.5	4	-	-	-	-
		Wet leaf	11,7	5.95	29	-	-	72.78	21
		Dry leaf	56.8	-	-	-	-	724.7	205
Meadow plants		Stalk	7.64	7.64	37	-	-	-	-
Tobac co	Zakatala-67	Leaf	0.69	-	-	-	-	8.92	3
		Root	4.05	-	-	2.2	26	14.82	5
		Stalk	9.85	-	-	-	-	125.73	43
	Trapezon d-15	Wet leaf	10.8	-	-	-	-	138.85	47
		Stalk	23.8	-	-	12.2	142	100.83	34
		Wet leaf	0.83	-	-	-	-	10.50	4
		Wet leaf	12.9	8.79	43	-	-	53.50	15
Meadow plants		leaf	26.9	26.96	867				

CONCLUSION:

The research link the regularity of the distribution of the natural radionuclides (U-238, Ra-226, Th-232, K-40) has been described due to the scheme: about soil ▯ tobacco ▯ stalk of tobacco ▯ leaves, in the northern region of Azerbaijan. The regularity of free radicals influence on the density of natural radioactive isotopes has been discovered in tobacco.

REFERENCES:

1. Aliyev G.A, Niyazov A.K. (1981). The results of studyinq the distribution of natural radionuclides in soil. 6[th]deleg. Cong of All union society of soil scientists.
2. Aliyev D.A, Abdullayev M.A. (1996). Artificial and natural radionuclides in soil-vegetable cover of Azerbaijan. M.
3. Aliyev D.A., Ablyllayev M.A.,Alexakhin R.M.(1988). Total rules of ^{90}Sr and ^{137}Cs migration in soil-vegetable cover of Azerbaijan.
4. Abdullayev M.A., Aliyev J.A. (1987). Longliving natural radionuclides in soil plants of the Azerbaijan SSR.In book: Proccedings - Budapest, vol.2.
5. Abdullayev M.A., Aliyev J.A. (1998). Migration of artificial and natural radionuclides in the system of soil – plant, Baku [Elm].

PREVENTION OF ACCIDENTAL EXPOSURES TO RADIODIAGNOSTICS AND RADIOTHERAPY PATIENTS
Radiation Safety Aspects

MOHAMMED K. ZAIDI AND THOMAS F. GESELL
Radiological and Environmental Sciences Laboratory, Idaho Falls,
and Idaho State University, Pocatello, ID. USA.

Corresponding author: zaidimk@id.doe.gov

ABSTRACT:

New developments in the field of radiological equipment and prevention of accidental exposures to patients are being presented. The rapidly developing technologies provide valuable medical benefits to suffering patients. The consequent incorrect use of these machines increases the mistakes and incorrect doses. Clinical examinations should be performed with the lowest achievable radiation dose to the patient, consistent with diagnostic quality. A quality assurance program, implemented by dedicated people, could be a key to improved radiographic diagnoses and the prevention of accidental exposures in radiation therapy.

Keywords: medical radiation, radiation safety, x-rays, cancer.

INTRODUCTION:

Medical applications of ionizing radiations are the largest contributor (about 95%) to average human exposures from man-made sources of radiation [1]. Even though, with the advancement of technology, new computerized procedures have been introduced which makes the work faster and potentially safer, radiation exposures are still of concern.

Ionizing radiations are used worldwide for medical diagnosis and therapy. They are generally confined to anatomical regions of clinical interest and provide benefit to the examined or treated patient. Today Radiology is divided into specialties Diagnostic Radiology and Radiation Therapy or Radiation Oncology. The physicians (radiologists), technologists and Medical Physicists specialize in only one of these areas. As such this paper will describe each area separately.

Exposures in Diagnostic Radiology are characterized by low doses to individual patients. Radiological risks associated with diagnostic procedures are typically low but good radiation protection practice requires that the patient's exposure be so managed that it is no higher than needed to obtain the required diagnostic information. Interventional radiography is a diagnostic procedure. It is used for example to image cardiac catheterization procedures. However, it can also result in relatively high doses. In spite of this knowledge, there have been recent reports of severe radiation related accidents due to poor practice of some interventional techniques discussed in ICRP-86.

M.K. Zaidi and I. Mustafaev (eds.), Radiation Safety Problems in the Caspian Region, 107-120.
© 2004 *Kluwer Academic Publishers. Printed in the Netherlands.*

Radiation Therapy doses, on the other hand, are designed to provide a high dose to the tumor volume while sparing the organs at risk and other healthy tissues. Consequences of accidental exposures in radiotherapy can be very serious and even fatal, as cases reported in section 4 of this paper. Quality assurance (QA) is required at all steps of planning and delivering dose so that the target volume receives the requisite dose while organs at risk and healthy tissues are spared to the extent feasible.

In both disciplines, in order to achieve adequate protection from ionizing radiation there has to be collaborative effort and teamwork, where professionals involved in the healthcare program have a role to play. Implementation of appropriate quality assurance programs, safe practice culture, continuing education and training of concerned staff are essential to achieve the objectives of radiation protection together with high quality therapy and diagnosis.

Many healthcare professionals receive radiation exposure during the course of their professional activities. The International Commission on Radiological Protection (ICRP) [2] provides numerical upper limits of dose for the control of occupational exposure for radiation workers. But more than that, optimization of work practice and dose delivery techniques are the main tools for efficient dose reduction.

Dr. R. N. Bryan, President of the Radiological Society of North America (RSNA) said that a change is coming in the world of radiology, as a digital revolution [3]. The advantages of digital imaging and computer controlled radiation therapy are overwhelming. For diagnostic radiology it means there will be immediate access to images and ancillary data. In radiation therapy it means that very complex radiation patterns can be generated to accurately treat cancer. This computer-generated technology will have a lasting impact. However, with all these advancements in technology, the mistakes are still made and result in poor diagnosis and treatment.

The radiology departments in the USA spend millions of dollars building and renovating facilities and purchasing new equipment but relatively little time or effort is made in developing a better understanding of the people who work in the radiology departments. There is a need to have adequate training to operate these sophisticated machines and save people's lives [4].

There is also a great need for more technical staff due to the increased use of more sophisticated technology. In the last few years' radiation therapy has added three dimensional (3D) conformal therapy and Intensity Modulated Radiation Therapy (IMRT). Highly technical and well-educated staff must operate these technologies. On average, across the USA alone, there is a shortage of radiation therapists, *i.e.* we need 1800 additional therapists. It will take five to ten years to resolve the problem despite notable gains in graduate education [5]. Areas of diagnostic radiology, such as nuclear medicine, have similar shortages with the advent of PET scanning.

DIAGNOSTIC RADIOLOGY:

There are some excellent machines being used in medical diagnostic practice, as discussed below, and research is ongoing to increase accuracy, reduce imaging time, and impart less dose to the patient.

Diagnostic Radiology helps us to examine the patient for broken bones, ulcers, tumors, diseases or malfunctions of various organs by producing diagnostic images, for the physician's interpretation. In many instances, the radiology technician works independently, while for some advanced procedures the radiologist and the technicians work together as a team to take more information and record better images. Proper filtration, optimal x-ray tube voltages, and high-speed image receptor systems have helped to reduce the dose to patient. The advent of rare earth screen/film systems in the 1990's helped in reducing the dose to the patient. Today the most common film/screen system has a 400 speed, reducing the patient dose by a factor of 4 from years ago. Film is slowly being replaced by digital image technology. Two types of flat plate imaging have evolved: Digital Radiography (DR) and Computed Radiography (CR). Both technologies have reduced spatial resolution, but increased contrast and latitude.

Digital Radiography (DR) was first demonstrated in 1995 and has brought greater efficiency and productivity to general radiography procedures. A DR equipped exam room can image more patients per day than a film-based exam room. In some institutions it is so effective it has replaced two x-ray systems with a single DR-based exam room. The biggest negative for DR is its initial cost. DR systems add at least $150,000USD for each plate installed. The ability to produce digital images streamlines workflow and enhances department productivity. To utilize a DR system a completely new digital radiographic data acquisition was developed including, digital display, real-time or near real-time monitoring, and digital post-processing. DR radiographs can be printed on film or can be read on a digital softcopy display terminal. These digital imaging systems are based on linear diode arrays and area diode arrays that are optically coupled to x-ray-sensitive scintillation material. Large objects can be imaged in near real time with extended linear arrays, and in real time with lens focused area arrays. Relatively high spatial resolution and high contrast that approach capabilities of film have been obtained. DR takes less time and produces better quality images and reduced radiation dose to the patient [6] DR systems have a relative speed approaching film/screen systems. In the film-less environment it is emerging as one of the best systems because it saves time and money by not using film. On the other hand, it is a very costly undertaking, as it needs a hefty capital investment. DR might soon become a standard of care [7].

Computed Radiography is the most predominate digital imaging technology today. It has been in use for many years but recent innovations have significantly enhanced both image quality and productivity. CR is popular because it does not require replacing existing x-ray equipment with new technology. Film cassettes are replaced with digital plates that store the x-ray image. After exposure special readers process CR plates. Once the images are read they are handled much like the DR images. The images can be sent to a softcopy display for reading or printed for interpretation. One CR reader can service several x-rays rooms. As such it is very affordable. Its major shortcoming is that it is equivalent to a 200-speed film/screen system. To convert from film/screen to CR requires the patient dose to double for each radiograph taken.

Increase in computer technology has both allowed and encouraged digital imaging to grow rapidly. With out affordable computer technology neither DR nor CR would be affordable. At the same time there must be a way to read and store these digital images. Picture Archiving Communication System (PACS) is the acquisition of the digital radiological images, their storage, transmission and display. PACS includes workstations for interpreting and reviewing images, archive stations for the storage of the images, and a computer network for transmitting image data files around the institution [8]. It has virtually eliminated retakes and improved the consistency of radiological images. The ability to adjust contrast and brightness on a diagnostic workstation helps to detect clinically significant and much more important information and has enhanced the patient care and productivity. By eliminating retakes through digital image processing patient doses have been reduced. Portable x-rays have one of the highest retake rates of all radiographic procedures. Digital image processing have significantly reduced the rate of retakes from portables by post processing the images and correcting for poor techniques used to take the radiographs.

To over see digital image technology the American College of Radiology (ACR) has set up an Imaging Network (ACRIN). It was established in 1999 and is a cooperative group that is dedicated to the performance of trails of medical imaging technologies as related to cancer diagnosis and treatment [9].

Digital fluoroscopy has streamlined workflow, expanded the type of procedures that can now be done in radiology and provided enhance department productivity but has not reduced patient dose. In fact patient dose has increased with the introduction of new interventional procedures. These new procedures significantly add to the life saving technology available in medicine, but require lots of fluoroscopy time, often long high dose rate image acquisitions.

Computerized Tomography (CT) was introduced in 1972. From that date forward CT examinations are increasing in number. Today 25% of all radiology procedures done in the hospital are CT's and account for about 75% of the diagnostic medical exposure [10]. The technical and clinical developments in CT have not generally led to reduction in patient dose per examination. The technological developments have significantly improved image quality and speed of exam. However, there are basic limitations that restrict dose reduction. Enough x-rays must be transmitted through the patient to form an acceptable image. Technology has not yet been able to reduce the number of x-rays needed and still generate a diagnostic quality image.

Some of the new CT scanners have automatic exposure control and this is helpful in dose reduction. However, this has yet to be proven. Pediatric CT exposures have been reported to be excessive because the same techniques used for adults are used for infants and juvenals. There have been several articles published recently suggesting techniques that reduce pediatric techniques thereby reducing patient doses while maintaining good image quality. New CT scanners have special pediatric techniques that require much less radiation exposure.

The reason CT scanning is increasing at such a rapid rate is that the new technology offers such new diagnostic tools to all physicians. CT can be used for detection of cancers that are as little as few millimeters in size [11]. Helical or corkscrew CT scanning has reduced scan time to less than 100 milli-seconds/slice. Slice thickness can vary from 1mm to 10mm. Organs on the order of 20-30 cm can be scanned in under 30 seconds. The 4D CT (temporal) was introduced at the ASTRO 2002 annual meeting where they have incorporated respiratory gating. Such fast scanning now allows new cardiac studies to be done on routine CT hardware.

The newest systems incorporate 16-slice technology (i.e. 16 slices per scan rotation). This technology provides unparalleled scan acquisition speed and resolution. This new acquisition scheme may reduce patient radiation dose [12]. At the 2002 RSNA meeting all major CT manufactures were showing research cone beam CT scanners with 256 simultaneous slice acquisitions. This technology should be commercially available within the next 5 years.

In the last two years mammography has not seen much in the way of new technology. GE introduced digital mammography followed by other vendors. However, this is still very expensive, as such has not expanded rapidly. Fuji should be FDA approval for its mammography CR system before the end of this year. This system is more cost effective and may allow more users to digitize mammography.

Yet, the demand for mammography services is increasing tremendously. Over the last few years there has been tremendous pressure from the US FDA and other regulatory bodies to control the quality of mammography. In parallel to this pressure is the low reimbursement for routine mammography screening. This has created a great shortage of radiologists specializing in breast imaging. The practice of mammography has become financially tenuous. Radiologists are moving where there is more money and less risk of regulatory sanctions or litigation. In a recent survey, the majority of residents did not want to interpret mammograms in their future practice because they are afraid of medical malpractice liability related to the interpretation of mammograms. Also low reimbursement is having a negative impact on both academic and community practices. The current shortage of radiologists in the US and Canada also has a negative impact on recruitment into mammography [13]. There is also controversy on the value of screening with mammography. A Danish report has claimed that the screening trials were biostatistically flawed and invalid [14]. A Canadian report also says that mammography screening has not reduced the death rate in breast cancer patients [15]. However, American physicians took issue with these reports at the ASTRO 2002 annual meeting indicating that the women who receive annual mammograms are more likely to have their breast cancer detected at its earliest and most curable stage.

Nuclear Medicine is often a subspecialty of diagnostic radiology. It aids in the diagnostic process by producing images or dynamic studies of the function and structure of the patient's body organs by imaging radioactive pharmaceuticals localized in the body. Digital gamma cameras and scanning machines are providing the state of the art for imaging the distribution of radionuclides in diagnostic nuclear medicine procedures. Technicium-99m (Tc-99m) is often the isotope of choice because it has a convenient half-life of 6 hours, emits gamma rays at energies optimal for imaging and binds well with the radiopharmaceuticals.

The newest nuclear medicine technology is called Positron Emission Tomography (PET). It was developed from its initial use for brain imaging in the 1980s to a main role in oncology and other medical specialties, with 350,000-plus scans performed in 2002. The scanner consists of an array of detectors that surrounds the patient. The scanner measures the pairs of high-energy 511 keV annihilation photons traveling outward in opposite directions produced from Fluorine-18 (F-18) in a fluorodeoxyglucose solution (FDG-18), which is injected into the patient. PET measures the amount of metabolic activity at a site in the body and a computer reassembles the signals into images. Cancer cells have higher metabolic rates than normal cells, and show up as denser areas on a PET scan. PET is useful in diagnosing certain cardiovascular and neurological diseases because it highlights areas with increased, diminished or no metabolic activity, thereby pinpointing problems. PET appears to be a more accurate indicator of treatment response than CT.

PET is now the initial diagnostic study of choice for determining myocardial viability, epilepsy patients, and mild to moderate Alzheimer disease, breast cancer at its earliest and most treatable stage, and treatment planning. The cardiology applications of PET address issues related to the efficient diagnosis and treatment of an increasing incidence of heart disease, the rising cost of cardiac care, and shortage of medical professional in this clinical segment. PET and functional MRI may help locate the microscopic abnormalities in brain structure and locate functional responsibility for many mental disorders and waves patterns that appear unique to patients with depression [16].

PET used in conjunction with CT scanners (PET/CT) combines anatomic and physiological data. It is very helpful in oncological imaging, especially in lung, colorectal, head and neck, and esophageal cancers as well as melanoma and lymphoma. It can improve staging and monitoring of disease in the case of multiple melanomas. PET/CT offers speed, image quality and patient acceptance.

With all of the good news from PET, the technology has a few safety concerns. Tc-99m based radiopharmaceuticals require only a small amount of shielding (usually about 0.3 mm of lead) to protect hospital staff from the 140 keV gamma rays. F-18 in the form of FDG is used in PET imaging. It is a positron emitter and produces two 511 keV photons. A room used for PET imaging typically requires about 4 mm of lead [17]. Most nuclear medicine rooms do not have this much shielding. Even rooms designed for conventional CT require more shielding before PET studies can be done in the room.

Another nuclear medicine study utilizing Single Photon Emission Computerized Tomography (SPECT) is still very active. Most SPECT studies are done to study cardiac function. The new SPECT cameras utilize 2 or 3 detectors to speed up the studies. Multiple detectors double or triple the tomographic slice sensitivity, allow 3D image reconstruction and perform whole body and spot imaging of the anterior and posterior views simultaneously.

RADIATION THERAPY:

Radiation Therapy is the discipline used to treat cancer patients. To treat cancer patients, large doses of radiation are delivered to the cancer cells while minimizing the dose to the normal cells surrounding the cancer. Unlike diagnostic radiology where the goal is to provide good images, while minimizing the radiation dose, in radiation therapy it is to precisely deliver a high dose, with little concern for imaging. The emphasis is on precise delivery of the radiation beam. Since these doses are lethal, small errors in delivery can cause catastrophic results.

Most patients are treated with external beam radiation therapy. In the past it has been known as teletherapy. External beam radiation therapy is the application of radiation in which the patient is placed directly into an intense radiation beam. The use of Co-60 teletherapy machines is declining, and electron accelerators producing x-rays and electrons are being used increasingly.

Accelerators provide flexibility in the choice of radiation energies and fewer regulatory problems because they contain no radionuclides and are completely safe when turned off. Accelerator technology also allows the use of respiratory gating.

High energy linear accelerators can have induced radioactivity in the components and air around them when operating over 10 MV. The radiation exposure potential from induced radioactivity is usually minimal to staff immediately entering the accelerator room after a patient's treatment because of rapidly decaying individual nuclides. There is normally insignificant direct radiation exposure, contamination, and inhalation hazard for the accelerator worker, the public and the environment at the facility.

In order to deliver lethal radiation doses to the cancer it must be fractionated. External beam treatment is usually delivered in 20 to 40 fractions or sessions. Radiation therapists or technologists assume direct responsibility for the well being of the patient preparatory to, during and following the delivery of daily treatment. The radiation therapy physicist is responsible for the treatment planning process. This includes quantitative analysis of the energy distribution of ionizing radiation used in radiation therapy. Accurate planning is needed to ensure a therapeutic prescription of radiation is applied to the patient in a safe manner. For a successful coarse of radiation therapy the dose delivered to the patient must be known to within $\pm5\%$.

Until recently the radiation beams used in external beam therapy were uniform in intensity and blocked to the shape of the tumor. Intensity modulated radiation therapy (IMRT) was introduced in 1990s. In this technology the intensity of the beam is modulated to generate the optimum dose distribution in the tumor. This requires a linear accelerator that is dynamically controlled. Its rate of adoption in the US health care market place was been slow but over the last few years increased dramatically. A multimillion-dollar investment is needed for a digitally controlled linear accelerators and treatment planning software systems. The use of IMRT increases the workload approximately by a factor of two. The advent of IMRT has created an acute shortage of trained medical physicists and radiation therapists [18]. In order to use existing linear accelerators it is often necessary to add shielding to the treatment rooms. The new shielding is needed to compensate for the increased head leakage during the long and inefficient treatment times.

A new specialty in radiation therapy had developed over the last 10-15 years. This is called Stereotactic Radiosurgery. It started as a way of treating brain disorders with a precise delivery of a single high dose of radiation. It was limited to head and neck lesions as these areas can be readily immobilized with skeletal fixation devices. The radiation field is treated by sharply defined radiation beams delivered to specific area of the brain to treat abnormalities, tumors or other functional disorders. It is now also being used to treat spine lesions, lungs and liver diseases.

There are now three basic forms of instruments to provide this treatment: Particle beams, Co-60 beams, and linear accelerators producing x-rays [19].

The Gammaknife was the first instrument used to deliver intense gamma rays onto a target, such as trigeminal nerve. It uses many intersecting beams of gamma rays to destroy a tumor or vascular malformation within the head. Local anesthesia is administered and a stereotactic frame is attached to the head. The head is imaged using a MRI or a CT scanner while the patient wears the frame. The imaging information is used to precisely direct the gamma radiation to the desired location. It is used for the treatment of brain disorders. There are more than 120 units worldwide, with which more than 20,000 patients each year are treated. The installation cost is less than that of linear accelerator as it requires very good shielding [20].

The other major mode of radiation therapy is to deliver the dose to the tumor by placing radioactive material inside the tumor this is called Brachytherapy. In this modality the radioactive sources are placed in direct contact with tissue to deliver very high dose.

Brachytherapy is often used to treat early stage (localized) prostate cancer and offers a number of advantages compared to alternative treatments. Brachytherapy of the prostate is usually done by implanting I-125 or Pd-109 seeds into the prostate. The treatment is done on an outpatient basis, allows rapid patient recovery, and has reduced side effects such as impotency and incontinence. Brachytherapy typically delivers a higher dose to the prostate than external beam techniques, and is an excellent alternative to surgery in patients unwilling or unsuitable for surgery [21]. These seeds are very low energy and produce minimal radiation risk to family and friends.

Low-dose-rate (LDR) brachytherapy may be used alone for early-stage cancers in accessible locations such as the oral cavity, lip, or nose and combined with external radiation treatments for larger ones. In these cases, interstitial implants are used to deliver a "boost dose" to the bulkiest tumor area. Implantation takes place in the hospital, with a brief period of anesthesia, and necessitates a hospital stay of several days. The implant is usually performed by placing hollow, stainless steel inserts directly into the tumor and then "afterloading" Ir-192 in the form of "needles". Afterloading consists of placing the radioactive material into the previously implanted inserts. Following placement, x-rays confirm the exact location of the needles, and dose calculations are rechecked. The results of brachytherapy are excellent for early-stage cancers of the head and neck [22].

High dose rate (HDR) after loading machines are used to treat cancers of the vagina, cervix, endometrium (the lining of the uterus), prostate, bladder and rectum cancer by placing radioactive pellets (commonly HDR sources) into soft tissue using a computer-controlled machine. A high dose of radiation is quickly given to the cancer with relative sparing of adjacent normal tissues. The sources are then quickly withdrawn when the treatment time is complete.

Intravascular brachytherapy (IVBT) is one of the only radiation therapy procedures not used to treat cancer. It is used in prevention of restenosis following treatment to remove arterial blockage. In one method, an Ir-92 gamma radiation source is used to deliver the treatment dose via a centered segmented end-lumen balloon catheter using a HDR afterloader. Efficacy has been demonstrated with three years of clinical trials, and IVBT may become a standard for care for preventing restenosis with potentially minimal side effects [23]. Beta particle emitters such as Sr-90 have been found to be similarly successful in reducing restenosis. Less unwanted radiation is delivered to the patient, and occupational exposures of staff are reduced.

Radioactive material can be used to treat cancer by injecting it directly into the patient. Several new types of treatment have been developed lately. One of the newer isotopes is Yttrium Y-90. It is a pure beta emitter with a maximum energy of 2.3 MeV, which has a tissue penetration of 2.5 mm. It is used in the form of microspheres to treat the cancerous modules in the liver. The microspheres are half the diameter of a human hair. Y-90 is a beta emitter and must be shielded with plastic instead of lead. The vial containing the microspheres is contained inside an acrylic container, which is in turn placed in a secondary acrylic container to contain any leakage. The beta particles produce some bremsstrahlung radiation, making it possible to image the spheres inside the body. There is almost no chance for any one to receive any radiation exposure because staff as the vial will be in an acrylic container during the angiographic procedure. Once the labeled microspheres are injected they will remain permanently fixed in position in the liver. The patient can then be released with minimal precautions as the body serves as the shield and the Y-90 decays to background after a few weeks. After the procedure, the room is surveyed for radioactivity and gowns, gloves, masks and booties are all collected in a radiation waste bag and monitored. The patient should be cautioned not go very close to young children for a few days [24, 25].

Radioactive iodine (I-131) is used to treat thyroid cancer patients. Single doses of up 250 mCi of I-131 administered with a cumulative dose up to 1500 mCi. There can be serious consequences if the wrong dose is delivered, so two individuals should do calculations - one done by the physician and the other done by the physicist and then compared. One should not simply check the other because there is a possibility of not detecting the error [26]. The I-131 may be administered by having the patient drink the liquid using a straw while leaning head towards the bottle or it may be administered in a capsule. The safety measures will change when encapsulated radioiodine is used in place of liquid. The bioassays are no longer necessary for the technologists who administer patient dose [27].

CASE HISTORY OF SOME ACCIDENTS:

Some of the major accidents associated with the uses of radiation in medicine are discussed in ICRP Publication 86, 2000, and are posted on the web page www.icrp.org/ICRP_86_RT_accidents.pps. Other accidents that have happened in the USA are listed on www.reports.eh.doe.gov/cas/accidents <http://www.reports.eh.doe.gov/cas/accidents>. A few of the illustrative accidents, reported in the above-mentioned web pages are discussed below:

TELETHERAPY (RADIONUCLIDE)

1974-76: The radioactive decay curve for a Co-60 teletherapy source was incorrectly plotted, so the treatment time was longer than appropriate - no beam measurements were done in 22 months and a total of 426 patients were treated-34% of them had severe complications. The U.S. Nuclear Regulatory Commission (NRC) requires that Co-60 teletherapy machine dose rates be checked each working day,

1982-90: In the United Kingdom (UK), a new Treatment Planning System (TPS) was acquired, but the parameters were never changed. The problem remained for 8 years while 1045 patients were treated with doses less than those prescribed. Four hundred and ninety two died because of recurrence, possibly due to underexposure.

1987-88: After a new Co-60 source was installed in a teletherapy machine, the Treatment Planning System (TPS) files were updated except a computer file, which was no longer in use and was not removed. Treatment planning was subsequently done incorrectly using the old computer file. There was no double or manual check for dose calculations, and 33 patients received 75% higher exposures than prescribed.

1996: In Costa Rica, a radioactive source was exchanged in a teletherapy machine. During the beam calibrations for the new source, a mistake in reading the timer lead to underestimation of the dose rate, and 60% more time than appropriate was used for subsequent patient treatments. The operators failed to note that the treatment times were too long for a new source with higher activity, and Independent calibrations were not done. One hundred and fifteen patients were adversely affected, and by two years after the overexposures 17 people had died as a result. One child affected by an overdose to brain and spinal cord lost his ability to speak and walk. A young woman became quadriplegic as a result of accidental overdose to the spinal cord.

2000: in Panama, a Treatment Planning System was to apply shielding corrections as the blocks were used. The computer erroneously calculated treatment times that were double the normal ones. The error affected 28 patients, five of whom died after a year from the overexposures

TELETHERAPY (ACCELERATOR):

1985-87: In USA and Canada, software from an older accelerator design was used for a new one. Six such accidental overexposures occurred in different hospitals and three patients died as a result.

1990: In Spain, an electron beam was restored but electron energy was misadjusted. The machine delivered 36 MeV electrons regardless of energy selected. 27 patients were affected with overdose and distorted dose distribution due to wrong energy. Seventeen of them died from accidental overexposures.

BRACHYTHERAPY

1992: In USA, a source detached from the driving mechanism of a high dose rate afterloading machine and remained inside the patient for several days. The patient subsequently died from the overexposure. The survey meter indicated that there was abnormal radiation present, but the operator erroneously decided that the source was in the shielded position.

2000: The U.S. NRC reported three misadministration that occurred at two different facilities during the conduct of intravascular brachytherapy (IVB) procedures because of improper dose calculation parameters being used. In one case the patient's coronary artery was treated with the IVB device. The intended dose was 8 gray (Gy) (800 rads). During a review of dosimetry and physician records, the licensee discovered that the diameter of the artery was used in the dose calculation, instead of the required radius of the artery. The licensee estimated that the dose to the outer portion of the patient's coronary artery was 14.6 Gy (1,460 rads) rather then the intended 8 Gy (800 rads). For each of the other two cases, a similar error resulted in estimated delivered doses of 12.5 Gy (1,250 rads) and 14.3 Gy (1,430 rads) for intended doses of 8 Gy (800 rads) [28].

2003: During a cardiac brachytherapy procedure conducted at a hospital there was a malfunction of the drive mechanism with an intravascular brachytherapy (IVB) device containing approximately 95 millicuries of P-32. The malfunction occurred during the treatment of the third of three patients. The first two treatments were completed without any problems. The treatment of the third patient was initiated with the dummy source reaching the proper dwell position successfully (confirmed visually via the fluoroscopy screen) and returning to the cartridge. The active source was then advanced into the catheter, but the licensee determined that the source movement light continued to blink well after the anticipated transit time. The licensee initiated a fluoroscopic view of the treatment site, and with no source in the field of view, the licensee assumed a machine malfunction had occurred and initiated emergency procedures. The licensee was unable to retract the source, so the attending physician removed the catheter and source from the patient and dropped them on the operating floor. After the power cord was removed from the wall receptacle, the source was retracted back to its shielded position.

ACCIDENTS UNRELATED TO PATIENTS:

Some accidents associated with medical radiation devices affected the public and other workers rather than patients. Abandoned radiotherapy sources caused public exposures and large-scale contamination in Mexico in 1984 and Brazil in 1988. In Mexico it resulted in 4000 people being exposed and contamination of metal used in the manufacturing of 30,000 table stands and 6000 tons of reinforcing rods. Eight hundred and fourteen buildings were partly or totally demolished as a result. In Brazil, 112,000 persons were monitored, 249 found to be contaminated, and four died as a result of a 1988 accident in which an abandoned teletherapy source was stolen, and the source removed and broken open by salvage yard personnel.

CLINICAL CONSEQUENCES:

The side effects of radiation therapy treatments are frequent, but usually minor and transient. The complications that are more severe and long lasting such as radiation myelitis are expected at very low frequency. In the case of underexposures, which may jeopardize the probability of tumor control, it often takes a long time to recognize the problem so a large number of patients may be involved. The impact of overdose is usually observed first in tissues with rapid cell turnover such as skin where effects such as necrosis increase in frequency and severity. The acute complications depend on total delivered dose, total duration and size and location of irradiated volume. There is very little correlation of early complications with fraction size and dose rate. Late complications are observed in tissues with slowly proliferating cells after about six months as they slowly progress to cause limitation of motion (e.g., hip) as a result of accidental overexposure. In serial organs such as spinal cord, intestine, and large arteries, a lesion of small volume irradiated above threshold may cause major incapacity and may end up in paralysis. In organs arranged in parallel such as lung and liver, severity is related to the tissue volume irradiated above threshold. The accidental overdose can be detected as early-enhanced reactions with clinical follow-up and weekly consultations. Some overdoses may cause late severe effects without abnormal early effects. In case of unusual reactions in one patient the other treatments during that interval should also be reviewed.

To help prevent these events, appropriate institutional arrangements should be required by regulations and compliance should be verified [29].

EMERGENCY PREPAREDNESS:

In order to prepare for radiation overexposure or radio-nuclide contamination incidents, the US Department of Energy Radiological Assistance Program (RAP) provides a flexible, around-the-clock radiological emergency response capability to federal agencies, state, tribal and local governments, and to private businesses or individuals for incidents involving radiological materials.

The Radiation Emergency Assistance Center (REACT) is sponsored by US Department of Energy and run by Oakridge National Laboratory, Oakridge, TN. This center provides training and plays a leadership role at the time of a radiological incidence.

CONCLUSIONS:

State-of-the-art QA programs for radiation therapy have evolved from equipment verification and prescription to delivery and post treatment follow-up. QA programs should cover the entire process. A good quality assurance program and a dedicated QA staff are critical to prevention of accidents under exposures or overexposures. Some of the operators have not reached the highest stage of safety awareness. Very little time or effort is devoted in developing a better understanding of the people who work in radiology departments.

RECOMMEDATIONS FOR PREVENTION OF ACCIDENTS:

A quality assurance (QA) program involving organization, education and training, acceptance testing and commissioning, follow-up on equipment faults, communication, patient identification and patient's charts etc should be established and maintained. The use of radiation sources and equipment should be governed by proper rules and procedures for personnel radiation protection, source storage, radioactive waste management and emergencies. Proper safety instructions should be documented and be given to patients, their attending relatives and the public.

Major accidents generally happen when the procedures are not followed and checks omitted. The QA process is not always properly followed. QA is crucial in accident prevention and should involve clinical, physical and safety components. Its implementation should include complex multi-professional teamwork, clear allocation of functions and responsibilities, functions and responsibilities understood, and number of qualified staff commensurate with workload. The three basic principles of radiation safety should always be remembered - time, distance and shielding.

Most important is training - personnel, including radiation oncologists, medical physicists, technologists and maintenance engineers need to be qualified and trained. Specific training should be given on procedures and responsibilities, everyone's role in the QA process, lessons learned from typical accidents with description of methods from prevention, additional training when new equipment and techniques are being introduced.

Acceptance testing should include tests of safety interlocks, verification of equipment specifications, as well as understanding and testing treatment-planning systems (TPS). Crosschecks and manual verification should be done routinely on TPS. Commissioning should include measuring and entering all basic data for future treatments into computer. Crosschecks and independent verification form a major part of accident prevention. If an equipment fault or malfunction has not been fully understood and corrected, there is a need for communication and follow-up with the manufacturers and this information and experience should be shared with others. In all there should be a need for a written communication policy including reporting of unusual equipment behavior, notification to the physicist and clearance before resuming treatments and reporting of unusual patient reactions. Effective patient identification procedures and treatment charts, double check of chart data at the beginning of treatment, before any changes in the course of treatment, and once a week at least during treatment. Adequate in-vivo dosimetry would prevent most accidental exposures. The initial beam calibration and follow-up, independent verification of calibrations, participation in dose quality audits should be the basis of a good quality program.

In brachytherapy the source activity, source positioning and source removal, dose calculations and treatment planning should always be double-checked. Patient monitoring is a very important part of this treatment. The operator should be trained and supervised by the medical physicist and the radiation oncologist. Trained staff should routinely conduct the general safety and software checks.

The Radiological Physics Center is a non-profit, NCI-funded resource to the radiation therapy and medical physics communities. This center should be used as a QA center to evaluate the calibration of all external beam machines and radioactive sources. Their web page is <http://rpc.mdanderson.org>. They can be contacted for more information and help.

Unneeded radioactive sources represent a potential liability to medical institutions and should be recycled or disposed of when feasible. The International Atomic Energy Agency (IAEA) is working diligently to account for orphan radioactive sources and may be able to provide assistance with disposition of unwanted sources [31]. To insure qualified personnel are handling radiation sources and equipment in some developing countries require certification and licensing. Radiological technologists and medical physicists are often licensed.

Acknowledgement: NATO provided full financial support to present this paper at the NATO ARW, Radiation Safety Problems in the Caspian Region, held in Baku, Azerbaijan in 2003.

REFERENCES:

1. UNSCEAR Report. (1996). Sources and Effects of Ionizing Radiation.
2. ICRP. (2000). Annals of the ICRP Publication 87 - Managing Patient Dose in CT.
3. Bryan, R.N. (2002). Digital Revolution-http://www.rsna.org/rsna/media/plenary.html.
4. Gunderman, RB. (2003). Understanding & Enhancing Work Perf. Rad. 227 (3). 623-26.
5. Cassady, JR. (2003). ASTRO Workforce Comm. report. IJROBP 56 (2). 309-18.
6. Smeets, P. (2002). Diagnostic X-ray. Advance IOA 12 (8). 13.
7. http://www.llnl.gov/IPandC/op96/09/9d-dig.html.
8. Qi. H. e.a. (1999). Content-based image retrieval - J. Dig. Imaging 12 (2) Suppl 1 81-82.
9. Hillman, BJ. e.a. (2003). ACR Network. Radiology 227. 631-32.
10. Wiest PW, Locken JA, Heintz PH, Mettler FA Jr., CT scanning: a major source of radiation exposure., Seminars in Ultrasound, CT and MR. 2002 Oct;23(5):402-10.
11. Pannu, H. (2002). Computerized Tomography. Advance, IOA 12 (8).13.
12. Fishman, EK. (2002). Computerized Tomography. Advance IOA 12 (12).
13. Bassett, LW. e.a. (2003). Survey of Radiology Residents. Radiol. 227 (3). 862-68.
14. Olson, O. (2001). Cochrane review with mammography. Lancet 358. 1340-2.
15. Miller, AB., e.a. (2002). The Canadian Study-1. A.Int.Med. 137 (5). 305-12.
16. Mayberg, HS. e.a. (2002). The Functional Neuroanatomy of the Placebo Effect. Am J. Psychiatry 159 (5). 728-737.
17. Courtney, JC. e.a. (2001). Photon Shielding. Health Phys 81 (Suppl). S24-28.
18. Mechalakos, J. (2002). Shielding Factors L. Accelerators. H.Phys 83. S65-67.
19. Vogel, WJ. (2003). Stereotactic Radiosurgery. Advance IRT 16 (4). 30.
20. Long, S. (2002). Cancer Treatment. 15 (26). 14-15.
21. Culp, T. e.a. (2001). Radiation Protection for I-125. Health Phys. (Suppl.) 29-32.
22. Mendenhall, WM. e.a. (1991). Brachytherapy head and neck. Oncol. 5 (1). 87-93.
23. Waksman, R. (2000). Vascular Brachytherapy: J.Inv.Cardiol.12 (2s). 18-28A.
24. Martin, WH. (2003). Radiation Safety using Microspheres. AIRT 16 (4). 35.
25. X. Zhu. (2003). Radiation Safety Considerations. H.Phys. 85 (S1). S31-35.
26. (2000). Ann ICRP Publication 86 - Prevention of Accidental Exposures to Patients Undergoing Radiation Therapy. Page 15 and Achey, B. e.a. (2001). Some Experience with Treating Thyroid Cancer Patients. Health Phys. 80 (Suppl 2). S62-66.
27. Coan, AC. (2003). Safety Measures Radioiodine T. Doses. H.Phys. 84 (6). S260.
28. (2002). NRC INFORMATION NOTICE 2002-16.
29. Ortiz, P. (1998). Lessons learned accident in radiotherapy, IAEA TEDCOC-1045.
30. Radiological Control Standard DOE-STD Technical Standard 1098-99, July 1999.
31. Steinhaeusler, F. and Bunn, G. (2003). Protecting the Sources - Securing Radioactive Material and Strong radiation Sources. IAEA Bulletin 45 (1). 17-20.

NEW APPROACH TO USE OF POLYENE ANTIBIOTICS

V. KH. IBRAGIMOVA[1], I. N.ALIEVA[2], D. I. ALIEV[2]

[1]Institute of Radiation Problems, Azerbaijan National Academy of Sciences
[2]Baku State University, Z. Khalilov str., 23 AZ 1073/1
Baku, AZERBAIJAN.

Corresponding author: iradanur@mail.az

ABSTRACT:

This work is a review of the scientific literature dealing with physiochemical properties of dimethyl sulfoxide and polyene antibiotics. The results of studies on separate and combined effects of polyene antibiotics and dimethyl sulfoxide on membrane permeability are analyzed. Results of our experimental investigations on radioprotective and antitumor properties of dimethyl sulfoxide complexes with polyene antibiotics are presented and the prospects of their use in medicine are described.

Keywords: Dimethyl sulfoxide, polyene antibiotics, complex and radioprotector.

INTRODUCTION:

Dimethyl sulfoxide (DMSO) is obtained by oxidation of dimethyl sulfide [1]. The properties of DMSO such as amphiphily, polarity, high resorption rate and quick penetration into organs and tissues, ability to dissolve many organic substances to molecular form make it especially attractive as a companion to other biologically active compounds. DMSO influences the stability of the lipid matrix of cell membranes [2,3], enhances membrane permeability [4], exhibits cryoprotector activity that allows one to extend storage of different cells and tissues at temperatures below 0^0C [5], and influences cell adhesion and differentiation [6,7]. In addition DMSO inhibits erythrocyte aggregation, improves microcirculation processes, normalizes fibrin formation, enhances phagocytosis, and inhibits chemotaxis of neurophils. Many of the above-mentioned DMSO effects were found independently of each other. However, they are not sufficiently studied at the molecular level.

The biological activity of many organic compounds sharply increases after their dissolution in DMSO. The DMSO spectrum region is between 350 and 2200 nm and the high solubility in DMSO of some antibiotics and proteins can be used for investigation of their molecular structure and Physiochemical characteristics [1,8-10]. DMSO is the main solvent of membranoactive polyene antibiotics (PAs), such as amphotericin B, nystatin, mycoheptin, and levorin.

By their chemical nature PAs belong to the polyene macrolides group of [11,12]. The UV spectra of Pas indicate that the PA macrolaction ring includes a polyene chromophore with a conjugated double bond system. PAs also contain a hydrophilic chain consisting of several hydroxyl and carbonyl groups, an aminosugar mycosamine, and an aromatic ketone *p*-aminoacetophenone.

M.K. Zaidi and I. Mustafaev (eds.), Radiation Safety Problems in the Caspian Region, 121-128.
© 2004 *Kluwer Academic Publishers. Printed in the Netherlands.*

The biological activity of Pas depends on the unique sterol content of membranes in complex with which the antibiotics form in lipid membranes oligomeric ion channels selectively permeable for ions and organic compounds [13].

This work deals with comparative analysis of physiochemical characteristics of DMSO and PAs, an analysis of their separate and combined effects on membrane permeability, describes our experimental results on the physiochemical characteristics of PAs in membranes, radioprotective and antitumor properties of the DMSO-PA complex, and describes the prospects of its use in medicine.

MAIN PHYSICOCHEMICAL PROPERTIES OF DIMETHYL SULFOXIDE AND POLYENE ANTIBIOTICS:

DMSO is very soluble in water. It is a transparent and colorless liquid with specific odor and slightly bitter taste. The data on some physical characteristics of DMSO and water are given in Table 1.

Table1. Physical properties of DMSO and water

Physical properties	DMSO	Water
Mol. weight, g/mol	78.13	18.02
Density at 20^0C, g cm^{-3}	1.1014	0.9982
Melting point, ^{0}C	18.4	0.00
Boiling point, ^{0}C	189.0	100.00
Refractive index at 20^0C	1.4770	1.3329
Surface tension at 20^0C	46.2	72.75
Viscosity at 20^0C, 10^{-3} Pa s	2.20	1.002
Dielectric constant at 20^0C	48.9	80.20

A relatively high boiling point and high latent heat of evaporation (53 J/mol 25^0C) show that DMSO molecules are highly associated and strongly bound to each other [14]. DMSO has dipole moment, and owing to the dipole-dipole interaction the molecules oriented a polymeric chain.

Upon mixing DMSO with water, an endothermic reaction leading to a temperature increase by 30^0 C takes place, which is indicative of hydrogen bond between DMSO and water molecules [14]. DMSO is a polar solvent, and owing to formation of strong hydrogen bonds with water molecules it can perturb the structure of the latter (this follows from the value of negative excess heat -2.556 J/mol at 25^0C). The enthalpy in the binary system DMSO-water acquires minimal values in the DMSO concentration interval 25-35%. The investigation of NMR spectra of aqueous solution of DMSO has shown minimal mobility of molecules (rotational and translational) is observed at 30% DMSO in aqueous solution [15]. The viscosity of the DMSO-water system reaches its maximum at the same DMSO concentrations. The results obtained indicate formation of DMSO-H_2O complex with a stoichiometric ratio 1:2. The study of surface activity of DMSO solution has shown that with increasing quantity of DMSO is well adsorbed on the surface of water. The surface tension of pure DMSO is approximately 60% lower than that of pure water (Table 1). The study of the radial distribution function by x-ray scattering has shown addition of a modest amount of DMSO into water (2, 6, and 10% w/w), which increases the ordering of water structure [14]. It is supposed that DMSO induces cooperative orientation of water molecules in such a way that hydrogen bonds between molecules become more rigid [2].

DMSO, glycerol and ethylene glycol exhibit cryoprotective properties, but DMSO is the most efficient of them. The optimal active concentration of DMSO in aqueous solution range from 4.8% to 40% [16]. Experiments on pig tissues, single cell embryos, leukocytes, blood serum, and even on different insect show that the use of the DMSO-containing medium is able to protect these objects against freezing for a long time [17].

Polyene antibiotics is a yellow powder without a distinct melting temperature; they have amphoteric properties and upon ionization form a cation in acidic medium and an anion in alkaline medium. The biological activity of PA sharply increases after dissolution in DMSO and they are tenfold more efficient as compared with their water-solved forms [17]. The PA solubility in acetone, isopropanol, and dimethyl-formamide is lower than that in DMSO. The surface tension of pure DMSO and its aqueous solution decreases after addition of polyene antibiotics. It should be noted that water-soluble sodium salts of PA lower the surface tension of water by 25-30%. The surface tension of the DMSO-PA solution is half that of pure water. Upon addition of DMSO-PA to an aqueous medium, self-aggregate of PA are formed, and its molecules and frequency of formation increase as the PA and DMSO concentrations are raised [18-20].

THE EFFECT OF DMSO-PA COMPLEX ON MEMBRANE PERMEABILITY:

Introduction of DMSO into cell suspension or model membrane system influences their permeability. Membrane permeability for DMSO and some nonelectrolytes was studied in details on the muscle fiber membrane of giant sea lobsters. Comparative data on the permeability for some compounds are given in Table 2.

Table 2. Coefficient of permeability and coefficient of distribution for some compounds in the oil-water system

Compound	Mol. weight	Permeability coefficient, 10^{-5} cm/s	Distribution coefficient
Water	18.00	26.1	-
DMSO	78.13	7.43	0.0030
Acetylsalicylic acid	300.26	2.98	1.78
Urea	60.06	1.05	0.0015
Glycerol	92.09	0.80	-

The DMSO molecules are able to penetrate quickly through biological membrane and overcome tissue barriers, including human skin. Data of Table 2 show that the size of molecules does not determine their permeability coefficient of the former is lower. Distribution coefficient also do not allow one to predict the extent of permeability for molecules. The distribution coefficient of DMSO is twice that of urea, whereas their permeability coefficients differ approximately seven-fold [1]. The DMSO penetration through membranes consisting of oxidized cholesterol exceeds that of urea approximately by 50-fold. These data show that the structure of lipid bilayer, along with that of penetrating molecules, is an important factor determining the permeation of water-dissolved compounds.

The DMSO molecules have a high degree of resorption, and since the dielectric permeability of DMSO is intermediate between those of water and lipids, DMSO is able to penetrate the hypodermic layers. This suggests that DMSO is capable of transporting drugs into these layers.

Polyene antibiotics practically do not penetrate poorly through bilayer membranes. The PA permeability coefficient for bilayer membranes consisting of brain phospholipids is 10^{-8}- 10^{-11} cm/s. The coefficient of PA distribution between membrane and solution depends on cholesterol concentration, and in the case of lipid-to-cholesterol ratio 20:1 it is of the order of 10^4–10^5. PA have an important peculiarity, namely, they sharply increase the membrane permeability for ions, water, nonelectrolytes, and organic compounds, when they are located at both sides of the membrane. The membrane conductivity increases in proportion to the 4th-15th power of PA concentrations, and this degree depends on the molecular structure of PA that forms channels. Under certain conditions, such as pH 3.0 and having the phospholipid concentration in the membrane-forming solution (to 10 mg/ml n-heptane), PA can efficiently increase the conductivity from one side of the membrane. In this case the membrane conductivity appears to be proportional to the 3rd-4th power of amphotericin B concentration. A sharp increase in membrane conductivity depending on PA concentration allowed us to suppose that ionic permeability is due to formation of an oligomeric structure in the membranes of polyene channels. The system responsible for selective permeability of membranes is localized in the hydrophilic chain of PA molecules, which consists of their hydroxyl and carbonyl groups. Polyene antibiotics may form hybrid channels upon introduction of different PA at the two sides of membrane.

The conductivity of such channels is intermediate between that of single ones formed by amphotericin B and levorin molecules. The conductivities of single amphotericin B, gramicidin, and alamethicin channels are in inverse proportion to the channel diameters. This suggests that the diffusion restrictions for penetration of antibiotics into the membrane and of ions into a pore are inessential. However, polyene channels are formed very slowly in membranes. According to [18], the time of establishment of stationary conductivity is -40 min; the disassembly of channels is a still slower process. The rate of channel formation grows as the DMSO and PA concentrations in aqueous solution and the sterol concentration in the membranes is increased. The formation of polyene channels may be connected with association of PA molecules in aqueous solutions into complexes, which may be integrated in blocks into membranes. For each PA these associates differ in size. After formation in membranes of a conducting oligomeric complex, a polyene channel has a tendency to independent disassembly within the membrane whereby it is converted to the nonconducting state. Channels can be disassembled in monomers or dimmers. The physicochemical properties of polyene channels were studied in water-salt solutions at DMSO concentrations of 0.1-1%. It was shown that DMSO enhances tenfold the efficiency of assembly of a polyene channel and stabilizes its work in the conducting state for a long time. The polar groups of PA molecules influence the duration of the channel conductivity. Alkylation or blocking of amine or carboxyl groups of PA molecules located at the channel entrance markedly decrease the average time of channel conductivity.

PA is involved in transfer of both ions and organic compounds [4] through membranes. An important problem of molecular biology is the search for systems of oligonucleotide transfer into mammalian cells, which are carried out in the framework of the Human Gene Therapy program. Positively charged PA screen the negative charge of nucleic acid molecules and make easier their transfer into a cell. As an example, we can point to the oligonucleotide transfer into mammalian cells using a positively charged derivative of amphotericin B, 3-dimethyl-aminopropyl amide.

RADIOPROTECTIVE PROPERTIES OF DIMETHYL SULFOXIDE AND POLYENE ANTIBIOTICS:

Cell membranes are primary target for ionizing radiation, in which the process of free radical formation is initiated. The radiation energy influences target molecules and thus induces formation of free radicals like formation of free radicals during water hydrolysis (the reaction rate constant is $7-10^9$ $M^{-1}s^{-1}$. DMSO in very low concentrations (0.001-0.01%) is able to efficiently block the formation of free radicals. It is usually considered as a genetically inactive compound able to protect DNA molecule against α and γ radiation. Owing to this, DMSO is often applied as a solvent of a number of compounds used in experiments on mutagenesis [1]. The mechanism of DMSO action is based on interaction with OH radicals and formation of a less reactive radical methanesulfonic acid (MSA).

$$H_3C - SO - CH_3 + OH \quad \rightarrow \quad CH_3 + H_3C - SO - OH$$
$$\text{DMSO} \qquad\qquad\qquad \text{MSA}$$

Our data shows that DMSO has certain radio protective properties. A single intraperitoneal injection of 0.01 ml DMSO to mice of 18-20g body weight 30 min. prior to irradiation increased the viability of animals. Thus, the survival of experimental animals by the 12th day after irradiation was 29%, whereas the animals of the control group died.

Radio protective properties of DMSO are enhanced upon its combination with PA. The latter was injected intraperitonally to mice from mother liquor (1mg antibiotic per 1ml DMSO) in a dosage of 0.01mg/kg animal body weight, 30 min. before irradiation to 7Gy. Among the PA studied, the most efficient were levorin and methylated levorin complexed with polyvinylpyrrodeolidone (PVP). The viability of the experimental animals by the 12th day after irradiation was 60%, and in control 20%. Like methylated levorin, PVP has no effect on the peripheral blood condition, but does not exhibit radio protective properties. The enhancement of radio protective properties of DMSO complexed with PA is, probably, due to the fact that owing to the existence in PA molecules of conjugated double bonds, a significant part of the energy of ionizing radiation can be absorbed by this system. An indirect support of this hypothesis is the inactivation by ionizing radiation of the polyene channel conductivity.

THE USE IN MEDICINE OF DIMETHYLSULFOXIDE COMPLEXED WITH POLYENE ANTIBIOTICS:

DMSO as aqueous solutions and gels is widely used in medical practice. The results of clinical investigations have shown that DMSO is able to dissolve gallstones and remove them from bile ducts. In total, 480 cholesterol-rich gallstones were isolated from gallbladders of 214 patients. The use of an EDTA- DMSO mixture makes it possible to clean the bile ducts of pigments and gallstones without surgery. It appeared that pure DMSO dissolves gallstones more efficiently than in a mixture with calcium-chelating agents. Intravenous injections of DMSO to 11 patients gave positive results upon treatment of amyloidosis, cytits, edema, and high intracranial pressure.

Clinico-experimental studies also showed that DMSO is efficient for local treatment of rheumatic diseases in the case of arthritis, tendovaginitis, bursitis, myositis, skin indurations, trophic ulcers, joint contractures (in the case of intra-articular injection of DMSO). It is assumed that the use of DMSO is the most efficient for treatment of secondary amyloidosis caused by rheumatic diseases, it may improve kidney function. It was shown that DMSO could penetrate into the articular cavity upon skin application. After application of 50% aqueous solution of DMSO, relatively high concentrations of the preparation (as compared with its level after 1 h) are preserved for a long time in blood serum and especially in urine, passing unchanged through kidneys.

The highest concentration of the orientation in urine were detected after its intake *per os*. Selective accumulation of DMSO in the salivary gland secretion served as the basis for local treatment of Sjogren's syndrome. A relatively high concentration of DMSO (15.9 ᴏɡ/ml) in the blood is observed only for 5h after application, and in 24 h it falls to 1.8 ᴏɡ. The use of a concentrated DMSO solution appeared to be the best nonsurgical method for treatment of fibrous contracture in patients with rheumatoid arthritis. DMSO is slightly toxic: according to the published data [1], its LD_{50} is 2.5-8.9 g/kg; according to our data, 15-20 g/kg.

PAs are widely used in medical practice for treatment of fungal infections. Polyenes are slightly toxic and exhibit chemotherapeutic activity against pathogenic yeast-like fungi (in particular, of Candida genus). The LD_{50} for amphotericin B, mycoheptin, and levorin upon oral intake is more than 9g/kg, 2600±500mg/kg, and 1100±200mg/kg; upon intraperitoneal injection it is 300-500/mg/kg, 9.1±10.6mg/kg, and 12.8±1.6mg/kg; and upon intravenous injection it is 4-9mg/kg, 1.04±0.3mg/kg, and 1.6±0.09mg/kg, respectively. These data were obtained after injection of aqueous suspension of PAs without DMSO. Table 3. DMSO distribution in biological fluids in different times after skin application of 1 g of a preparation

Biological fluids	Concentration, ᴏɡ/ml		
	in 1 h	in 5 h	in 24 h
Blood serum	4.5 ± 2.6	5.9 ± 2.2	1.8 ± 0.6
Saliva	17.2 ± 7.9	5.2 ± 1.7	1.9 ± 0.5
Urine	9.9 ± 1.3	11.2 ± 3.4	6.3 ± 2.7
Synovial fluid	-	1.6 ± 0.1	-

The sterol sensitivity of PA, their ability to selectively increase ionic permeability of cell membranes, as well as the substantially higher cholesterol concentration in membranes of tumor cells as the basis for PA application in chemotherapy of solid and ascites tumors. Mongrel albino rats with body weight of 110-130g were inoculated with cells of tumors of different histological structure such as an ovarian tumor, spindle cell sarcoma-45, Walker carcinosarcoma, Plisse lymphosarcoma, sarcoma M-1, Gueren carcinoma, and mouse Ehrlich ascites tumor. PA in DMSO (1 mg/ml) were injected (0.01 ml) into animals with ascites tumors in 24h and to those with solid tumors in 72h after inoculation, and injections continued for 10 days. PA enhanced inhibition of tumor growth as compared with control. Thus, an alkylated derivative of amphotericin B ethamphocin, butamphocin, and original levorin inhibit growth of Walker carcinoma by 71.8%, Plisse lymphosarcoma by 46.3%, Gueren carcinoma by 51.3%, mouse Ehrlich ascites tumor by 52.7%, and M-1 sarcoma by 27.3% compared with control. The PA studied had no noticeable effect on peripheral blood and weight of spleen. The extent of growth inhibition of different tumors probably depends on the molecular structure of the antibiotic as well as on the cell surface structure and membrane composition of particular strains of tumor cells.

When antibiotics are used, there emerges the problem concerning their quick elimination from the organism, because they are toxic in high concentration. Most PA is insoluble in water. A combined injection of DMSO-PA into the organism results in lowering the coefficient of PA distribution between the membrane and water. In this connection it was supposed that the use of PA with DMSO might result in increased efficiency of PA (the achievement of maximal treatment effect at minimal concentration of antibiotics), increased degree of resorption, and selective effect on pathogenic microorganisms. The selectivity of PA effect is due to the different sterol composition of fungal and host cells. The fungal cell membranes contain ergosterol, whereas the host cells contain cholesterol. Pathogenic fungal cells containing ergosterol are 10-100 times more sensitive to PA than cholesterol-containing host cells.

CONCLUSION:

The use of the PA-DMSO complex has broad prospects in medicine. The good solubility of the PA-DMSO complex in water is observed at low PA and DMSO concentration (1-4mg/ml DMSO). As the concentration increases to 5-10mg/ml in lowered solubility of the complex; formation of antibiotic micells and their precipitation are observed. This is probably due to the fact that at high PA concentration the molecules of water begin to compete with DMSO molecules for binding sites, displacing them from the complex, and stimulate formation of self-aggregates of PA molecules in aqueous solution. Higher PA concentration in DMSO (>10mg/ml) has good solubility of the complex in water ratio of 30:1. As this ratio grows, PA looses their biological activity and the size of aggregates of PA molecules sharply increases.

REFERENCES:

1. Yu, Z., Quinn, P. (1994). Dimethyl sulphoxide: a review of its applications in cell biology. Bioscience Reports, 14, 259-281.

2. Yu, Z., Quinn, P. (1998). The modulation of membrane structure and stability by dimethyl sulphoxide (review). Mol.Membrane Biology, 15, 59-68.

3. Yu, Z., Quinn, P. (1998). Solvation on effects of dimethyl sulphoxide on the structure of phosphor; ipid bilayers. Biophys.Chemistry, 70, 35-39.

4. Hsieh, D. (1994). Drug Permeation Enhancement, Theory and Applications. Dortmund.

5. Karlsson, J., Cravalho, E., Rinkes, I., Tompkins, R., Yarmush, M., Toner, M. (1993). Nucleation and growth of ice crystals inside cultured hepatocytes during freezing in the presence of dimethyl sulphoxide. Biophys.J. 65, 2524-2536.

6. Anchordoguy, T., Carpenter, J., Crowe, J., Crowe L. (1992). Temperature-dependent perturbation of phospholipids bilayers by dimethyl sulphoxide. Biochim. Biophys. Acta, 1104, 117-122.

7. Watson, R., Rotstein, O., Parodo, G., Bitar, R., Hackman, D., Marshall, J. (1997). Granulocytic differentiation of HL-60 cells results in spontaneous apoptosis mediated by increased caspase expression. FEBS Lett. 412, 603-609.

8. Grdadolnik, S., Mierke, D., Byk, G., Zeltser, I., Gilon, C., Kessler, H.J. (1994). Comparison of the conformation of active and nonactive backbone cyclic analogs of substance-P as a tool to elucidate features of the bioactive conformation NMR and molecular dynamics in DMSO and water. J.Med.Chem. 37, 2145-2152.

9. Johanneson, H., Denisov, V., Halle, B. (1997). Dimethyl sulfoxide binding to globular proteins: A nuclear magnetic relaxation dispersion study. Protein Sci. (USA), 1756-1763.

10. Verheyden, P., Franco, W., Pepermans, H., Van. Binst, G. (1990). Conformational study of a somatostatin analog in DMSO/water by 2D NMR. Biopolymers, 30, 855-860.

11. Brajtburg, J., Bolard, J. (1996). Carrier effects on biological activity of amphotericin B. Clin. Microbiol. Rev., 9, 512-531.

12. Cybulska, B., Bolard, J., Seksek, O., Czerwinski, A., Borowski, E. (1995). Identification of the structural elements of amphotericin B and other polyene macrolide antibiotics of the heptaene group influencing the ionic selectivity of the permeability pathways formed in the red cell membrane. Biochim. Biophys. Acta, 1240, 167-178.

13. Cotero B., Rebolledo-Antunez, S., Ortega-Blake, I. (1998). On the role of sterol in the formation of the amphotericin B channel. Biochim. Biophys. Acta. 1375, 43-51.

14. Vaisman, I., Berkowitz, M. (1992). Local structural order and molecular associations in water-DMSO mixtures. Molecular dynamics study. J.Am.Chem.Soc. 114, 7889-7896.

15. Yu., Z., Williams, W., Quinn, P. (1996). Archiv.Biochem.Biophys. 332, 187

16. Mahajan, R., Renapurkar, D. (1993). Cryopreservation of Angiostrongylus cantonensis third-stage larvae. J.Helminthol. 67, 233-237.

17. Ali, J., Shelton, J. (1993). Design of vitrification solutions for the cryopreservation of embryos. J. Reprod. Fertil. 99, 471-477.

18. Countinho, A., Prieto, M. (1995). Self-association of the polyene antibiotic nystatin in dipalmitoylphosphatidylcholine vesicles: Biophys. J., 69, 2541-2557.

19. Gaboriau, F., Cheron, M., Petit, C., Bolard, J. (1997). Heat-induced superaggregation of amphotericin. Antimicrobial Agents and Chemotherapy, Now., 2345-2351.

20. Gaboriau, F., Cheron, M., Leroy, L., Bolard, J. (1997). Physico-chemical properties of the heat-induced superaggregation of amphotericin B. Biophys.Chem. 66, 1-1.

RADIOLOGICAL IMPACT ON MAN AND THE ENVIRONMENT FROM THE OIL AND GAS INDUSTRY: RISK ASSESSMENT FOR THE CRITICAL GROUP

F. STEINHÄUSLER
Center for International Security and Cooperation
Stanford University, CA, USA.

Corresponding author: physik@sbg.ac.at

ABSTRACT:

Radiation exposure of workers in the oil-gas industry can occur by inhalation of high-levels of radon gas; increased gamma dose rates; increased gamma dose rate due to ^{226}Ra, ^{210}Pb, ^{228}Ra, and ^{228}Th. The waste also has elevated contents of long-lived radionuclides. Finally, the recycling of waste originating within the oil-gas industry can pose a contamination problem. The various exposure pathways and a regulatory framework are discussed with regard to their applicability to the conditions in the Caspian Region.

Keywords: Radon, radium, radiation protection, contamination, risk assessment, oil and gas.

INTRODUCTION:

Technologically Enhanced Natural Radiation (TENR) means radiation from naturally occurring isotopes to which exposure would not occur by (or would be increased by) some technological activity not expressly designed to produce radiation. TENR and the oil and gas industry had a strained relationship. For almost 30 years it has been discussed in the scientific community that some workers in the oil and gas industry are exposed to echnologically enhanced levels of natural radioactivity [1,2]. However, this industry was rather reluctant to acknowledge its employees as potentially occupationally exposed to radiation.

Since 1991 the situation has improved to some extent because the International Commission for Radiological Protection (ICRP) has recommended that the full system of radiation protection should also apply for these workers, provided the TENR exposure scenario in average dose values exceeds 1 mSv/yr [3]. The International Atomic Energy Agency (IAEA), recommending to its member states to adopt the ICRP recommendation [4], further expressed the concern over this issue. It is important to recognize that on the one hand, TENR exposure in terms of individual dose of a worker in the oil and gas industry is truly a global issue due to the global distribution of reserves. On the other hand, these figures clearly show that the impact on the collective dose is not uniform, i.e. the number of workers subject to TENR exposure is significantly higher in the Middle East and Central Asia as compared to the number of workers of all other regions combined. For example, the contribution to the total world oil production by the different regions is as follows (approximate figures):

M.K. Zaidi and I. Mustafaev (eds.), Radiation Safety Problems in the Caspian Region, 129-134.
© 2004 *Kluwer Academic Publishers. Printed in the Netherlands.*

□ Asia Pacific region: 10%
□ Europe: 10%
□ The Americas: 20%
□ Middle East, Central Asia 60%
   (incl. former Soviet Union):

In Azerbaijan, both oil and gas exploration is continuing to expand. From January to July 2003, the national State Oil Company (SOCAR) produced 5,196,000 tons of oil, exceeding the production of the same period in 2002 by 15,400 tons. The national gas industry produced in the same period of 2003 almost 3 billion m^3 of natural gas (an increase of 77.5 million m^3 over the forecast [5].

It is emphasized that in many countries the oil and gas industry represents a powerful concentration of capital and is frequently one of the main providers of a large number of jobs. Therefore this industry is able to exert also significant lobbying power at the political level. For example, in Brazil until recently a single company had the monopoly to extract oil in all of Brazil, making it the largest Brazilian commercial and industrial enterprise. Such a concentration of power can pose a significant hurdle in enforcing the implementation of any TENR-relevant legislation, as in the case of Brazil. The situation is still worse in many other countries with large oil and gas extraction industries, which have not even yet finished their internal discussion on how to adopt a common regulatory structure with regard to TENR. This is the current situation for: Argentina, Azerbaijan, Australia, Bolivia, China, Ecuador, India, Indonesia, Kazakhstan, Malaysia, Mexico, Saudi Arabia, United Arab Emirates, and Venezuela.

Exposure pathways

Exposure to the TENR in the oil-gas industry can occur as occupational exposure for the workers at various stages of the exploration processing, but also result in an environmental impact due to radioactive releases along various pathways. International databases on radionuclide concentrations in the environment of oil-gas extraction/processing facilities are scarce [6,7].

Incorporation

Workers employed in the vicinity of oil and gas-drilling operations will be exposed to radon gas (^{220}Rn, ^{222}Rn). The radon isotopes are contained in the natural gas, or they are dissolved in the different fluids resulting from the production process. Particularly at working areas adjacent to the well heads outdoor radon values have been found to average 500 Bq/m^3, at some processing plants radon flow reached up to 6.2 GBq per day. Also high radon concentration values have been detected at gas storage tanks and compressor facilities. The radon concentration (in Bq/kg) at different stages of production facilities can reach significant values [7]:

□ Demethaniser   438
□ Depropaniser reflux   633
□ Deethaniser vapour   1,393
□ Block oil   1,797

A second inhalation exposure pathway occurs by using the oil shale as a source of oil. Oil shale is a sedimentary rock, which contains kerogen. Oil shale is used to produce synthetic shale oil. Since oil shale contains on average about 250% of the ^{238}U content of coal, raw and spent shale particles released to the atmosphere during mining and processing can represent a significant pathway via inhalation and ingestion. Retort gases resulting from the processing of oil shale can have a ^{222}Rn concentration up to 18 kBq/m^3. Leaching of radionuclides from spent shale piles can contaminate the ground water, since the ^{226}Ra concentration can reach up to 42 kBq/m^3 [6].

Occupational radiation exposure for oil- and gas workers can also occur during maintenance of plant facilities. Maintenance workers opening valves or working with used pipes are likely to incorporate ^{226}Ra, ^{210}Pb, ^{228}Ra, and ^{228}Th from deposits formed on metal surfaces (scales). These scales result from the partial solubility of ^{226}Ra and ^{228}Ra in water. They can be found in the fluids present in the oil and gas reservoir, such as water, gas, oil or condensates. Subsequently the radium isotopes mix with barium, calcium and strontium, present in the rock formation of the deposit. Along the pathway from the oil and gas reservoir to the aboveground production facilities, the various treatment processes decrease the dissolution characteristics of the fluids for radium isotopes. This leads to the deposition of radium isotopes as radium carbonate or radium sulphate in the form of scales or sludge.

High dust levels occur in the breathing area of such workers sandblasting the inside of tanks, or brushing of surfaces for grinding the metal. The scales removed in this manner contain ^{226}Ra up to 1,000 kBq/kg (average: about 5 kBq/kg), and in addition (maximum values each in kBq/kg): ^{210}Pb (72), ^{228}Ra (360), ^{228}Th (360). ^{210}Pb exposure (and of its decay products) can also occur for workers whenever they touch internal surfaces (e.g., in pumps; [8]. These lead deposits can contain ^{210}Pb up to 3,000 kBq/kg [9].

Also non-occupational exposure via the incorporation of radionuclides can result due to practices associated indirectly with the oil and gas industry, such as:

◻ The use of sewage sludge from oil processing plants in agriculture. Since this waste can contain long-lived radionuclides (e.g. ^{226}Ra), this can lead to an undesirable contamination of agricultural products. Typically the total ^{226}Ra activity brought to the surface equals about 918 GBq/yr [10];

◻ Disposal of large amounts of contaminated wastes (scales, sludge) in so-called "lagoons" near platforms. This practice occurs frequently in remote desert areas, not necessarily subject to stringent environmental control. The average sludge and scale generated by an oil-producing well amounts to 2.25 t/yr [10];

◻ Spent oil-shale piles are frequently located in regions with increased precipitation (e.g., Scandinavia). Leaching of these piles has the potential for groundwater contamination, since the leachate can contain ^{226}Ra up to 42 Bq/l.

A special issue is the reuse of contaminated metal, which originated within the oil and gas industry. This has been found to represent a serious contamination problem for the operator of such a recycling facility. Dismantled metal equipment can contain: ^{226}Ra in the scales deposited on the inside of the tanks, and the inner surface of pipes, short-lived ^{222}Rn-decay products deposited on the inside of various equipment components; this itself leads eventually to the contamination with ^{210}Pb and the growth of [210]Po. Total activity values for the scale range typically from 500 to 800 kBq/kg [11].

In the first stage of the recycling the pipes are cleaned by high-pressure water jets (pressure up to 2,500 bar) to remove the scales from the inside of the pipes. Subsequently the metal is cut and molten in an electric furnace. The melting process separates the natural radionuclides and leads to an enrichment process in the slag and to a much lesser degree in dust (predominantly [226]Ra, which accumulates to about 98% in slag and only 2% in dust).[2]

External radiation: Workers on platforms or near rigs, exposed to increased gamma dose rate due to external radiation in the vicinity of vessels and tanks filled with brines raised to the surface. Both, workers as well as members of the public can be exposed to external radiation from the storage and transport of radioactive contaminated materials and waste products.

The magnitude of the environmental issue becomes apparent when taking into account that even in countries with a relatively small oil and gas exploration large amounts of brines are raised to the surface as an unwanted by-product; for example in Germany, almost 30 million t of brines annually [12]. The United Kingdom has about 80 offshore platforms, disposing routinely of radioactive TENR wastes from its production facilities [13].

Doses for critical groups

Critical groups are: members of the public, exposed to incorporation of radium via the food chain or through inhalation of radium containing dust, workers inhaling radon at the workplace, respectively incorporating radon, radium and lead. Actual release data are rare and therefore the doses have to be calculated, assuming different exposure scenarios as input data for dose models, such as BIOS (National Radiological Protection Board, Chilton, Didcot, United Kingdom).

Committed effective doses due to inhalation of radioactive aerosols (e.g., during the removal of scales in a tank) are summarized in Table 1. These values are based on the dose conversion coefficients as recommended in the EURATOM Basic Safety Standards [14]

Table 1: Committed effective doses due to inhalation and ingestion (mSv/g of incorporated material) for scales[1]

Radionuclide	Specific activity [kBq/kg]	Inhalation AMAD 1 αm	Ingestion
[226]Ra	500	1.6	0.1
[228]Ra	300	13.5	0.2
[210]Pb	300	1.2	0.3

---

[2] [210]Pb is gaseous at the typical operating temperatures up to 1 400 C and is accumulated to about 93% in the dust of the air cleaning system of the melter

[1] based on measurements of scales with the maximally observed activities in the soil and gas industry

It can be seen that the inhalation exposure exceeds in all cases the exposure due to ingestion by at least 400%. By comparison, external radiation due to radon and its decay products is rather low: assuming a ^{222}Rn flow of 6 GBq/d and an exposure for 20 h/week, this will result in an external gamma exposure of 0.15 mSv/yr.

The dose rate due to external radiation is a function of the activity concentration of the radium isotopes contained in the scales, the amount of scales, the irradiation geometry, and the shielding characteristics of the wall material contaminated on the inside with radium (e.g., of the contained filled with brines). The external dose rate is predominantly due to radium isotopes, since the ^{210}Pb deposits do not contribute significantly to the external irradiation due to the low energy of the gamma radiation (46 keV). The resulting dose rate values can reach up to several tens of ∝Sv/h [9]. If the employee is engaged in a dust-laden atmosphere due to the removal of scales, the additional respiratory tract-relevant doses are about 0.5 mSv/yr.
Assuming the lack of any precautionary actions during routine operations the following range of dose values can be expected [9]:

□ Maintenance: 0.03 to 0.5 mSv/yr
□ Revision: 0.3 to 7 mSv/yr

For comparison, the absorbed dose rate in air, inclusive cosmic and terrestrial exposure, ranges from 0.02 to 30 ∝Gy/h in areas of high natural background rates [15].

Regulatory control

In the European Union EURATOM recommends that all workplaces with a potential exposure to natural radioactivity must be evaluated. If a value of 1 mSv/yr is exceeded, appropriate measures have to be taken to control – and if necessary – to reduce the occupational radiation exposure.

In case an industrial activity results in NORM waste materials below 500 kBq/kg these materials are not subject to control, provided the above listed dose limit and nuclide-specific upper values are adhered to; for example, in case of ^{226}Ra this corresponds to a specific activity of 65 kBq/kg. In this situation there are no requirements for records to be kept of disposals made. If the limit of 500 kBq/kg is exceeded, a Member State can exempt this type of activity only if the individual dose of 10 ∝Sv, respectively the collective dose of 1 man-Sv is not exceeded. An alternative approach can be the demonstration that the exemption is indeed the optimum solution in terms of radiation protection, e.g. by applying an optimisation process.

The practical application of such regulatory approach for a smaller oil and gas producing country, such as Azerbaijan, is demonstrated in the case of UK offshore operations. For wastes resulting from the UK oil and gas exploration carried out under the terms of the Radioactive Substances Exemption Order, there is an exemption rule in place, which covers phosphatic substances, rare earths, etc. Otherwise, disposal of scale and other wastes is granted under the terms of the Radioactive Substances Act (1993), which indeed requires records of actual discharges made under the authorization to be archived.

CONCLUSIONS:

Globally there is still inadequate awareness in oil and gas industry about the necessity to address the issue of radiation protection for workers and controlling the discharge of radioactive wastes into the environment. Numerical data on the activity concentration and mass flow of radionuclides involved in the different stages of the oil and gas exploration and production are insufficient and subject to large uncertainties. This is specially the case for oil and gas exploration in developing countries.

Based on the few currently available data, critical groups in the workforce can be defined (e.g., workers engaged in revision) who may receive – under adverse conditions – significantly elevated doses exceeding the currently applicable dose limits for non-occupationally exposed persons. Therefore a worldwide initiative is recommended to improve the currently inadequate lack of statistically representative data for the radiation exposure of the workforce in order to a) identify critical groups, b) apply an adequate system of radiation protection for them. In addition, representative surveys of the environmental impact resulting from the waste disposal practices should be intensified to derive at cost-effective sustainable solutions with regard to an environmentally friendly waste management system.

Acknowledgement: NATO provided full financial support to present this paper at the NATO Advanced Research Workshop, Radiation Safety Problems in the Caspian Region, BAKU.

REFERENCES:

1. GESELL, T.F., (1975). Occupational rad. exposure due to Rn222.H.Physics 29 (5). 681.
2. STEINHÄUSLER, F. (1980). Assessment of the radiation. TAEC Tech.Jour.7 (2). 55-65
3. ICRP. (1991). Recommendations, Pub. no. 60, Pergamon Press, Oxford.
4. IAEA. (1996). International basic safety standards for protection. IAEA Safety Series 115.
5. IBRAHIM, K. (2003). Azer. Newsletter, Emb.Republic of Azerbaijan, Washington, D.C.
6. STEINHÄUSLER, F. (1990). Technologically enhanced natural radiation and related risks, Proc. Int. Conf. "High Levels of Natural Radiation", Ramsar, Iran, 163.
7. STEINHÄUSER, F., PASCHOA, A.S. AND ZAHAROWSKI, W. (2000) Radiological impact due to oil- and gas extraction and processing: Proc. IRPA Congress, Hiroshima.
8. ZAHAROWSKI, W. (1999). Naturally occurring radioactive.App.Rad.Isot. 48. 1391-96.
9. HOOGSTRAATE, H., VAN SONSBECK, R. (2002). Training on NORM: increasing awareness, reducing occupational dose, IAEA TECDOC no. 1271.
10. ENVIRONMENTAL PROTECTION AGENCY (EPA), Diffuse NORM. (1991). Waste characterisation; and preliminary risk assessment, USEPA, Air and Radiation (ANR-460).
11. SAPPOK, M., QUADE, U. and KREH, R. (1998). Naturally occurring radioactive material Proc. 30th Ann. "Fachverband für Strahlenschutz e.V.", Radioakt.Mensch.Umwelt, 1, Lindau.
12. KOLB, W.A. and WOIJCIK. (1985). Science of the Total Environment 45. 127-134.
13. WARNER JONES (2002). Proc. 7th Int. Natural Rad. Symp.P.No 0230, Rhodos (in press).
14. EUROPEAN UNION, Council Directive 96/29/Euratom. (1996). Basic safety standards for the protection of the health of workers.Official J.European Community, L 159, Vol 39.
15. UNSCEAR. (2000). Sources and Effects of Ionising Radiation, United Nations, NY, USA.

RADIATION-THERMAL PURIFICATION OF WASTER WATER FROM OIL POLLUTION

I. MUSTAFAEV, N. GULIYEVA, S. ALIYEV, I. MAMEDYAROVA
Institute of Radiation Problems, Azerbaijan National Academy of Sciences
31-a H. Javid ave Baku-370143, Azerbaijan.

Corresponding author: imustafaev@iatp.az

ABSTRACT:

During radiation-thermal decomposition of n-heptane, the main parameters of radiolysis change within the bounds: temperature 20-400°C, absorbed dose 0-10.8 kGy at dose rate 3.6 kGy/h. The correlation of n-heptane concentration and water steam is $[C_5H_{12}] / [H_2O] = (1-100)10^{-5}$. Total concentration of steam was about 10^{20} mol/ml. H_2, CO, CH_4, C_2H_4, C_2H_6, C_3H_8, C_3H_6, C_4H_8, hydrocarbons C_5, and C_6 are the products of decomposition. The changes of n-heptane concentration in the reactor were established and its decomposition at high temperatures in the irradiated mixture is observed and the mechanism of radiation thermal process in hydrocarbons-water system are discussed.

Keywords: water, heptane, radiolysis, purification, □-radiation

INTRODUCTION:

During the extraction, preparation, transportation and refining of oil the sewages containing oil contaminations are produced. The concentration of oil content in the water depends on used technology and may vary from a thousandths parts up to tens percents. There is a necessity of cleaning this pollution up to a permissible level. There are numerous methods (adsorption, mechanical, chemical and etc) of treating of waster water from oil contaminations [1-2]. Radiation-chemical method is one of the effective among the above mentioned methods [3-4]. Early it have been established that under conjugate effect of heat and radiation decomposition of raw hydrocarbon happens in chain regime [5]. In that case the radiation-chemical yield of water decomposition does not exceed 7-8 molec/100eV [6]. The experimental researches of radiation-chemical processes in water containing micro-impurity of oil were not conducted. At the same time for perspective evaluation of radiation-thermal purification method for refining sewages from oil contaminations it would be useful to have those data. The experimental researches of radiation-thermal conversion of hydrocarbon micro-impurities in water medium are adduced in this paper.

EXPERIMENTAL:

The experiments on study of the kinetics of radiation-thermal process were conducted on mixtures of water with n-heptane as a oil hydrocarbon model. The simulated mixtures have

M.K. Zaidi and I. Mustafaev (eds.), Radiation Safety Problems in the Caspian Region, 135-139.
© 2004 *Kluwer Academic Publishers. Printed in the Netherlands.*

prepared in vacuum equipment. The water vapor and n-heptane sequentially were frozen in the glass ampoule of 30 ml volume and were sealed. The total concentration of the mixture in the ampoule was 10^{20} mol/ml, and the components ratio changed in the limits $[C_7H_{16}]/[H_2O] = (1-100) \cdot 10^{-5}$. Irradiation of the ampoule has been carried out in ◻-radiation source of Co^{60}, by dose rate 3.6 kGy/h. In the all studied interval of temperature the n-heptane was in a vapor phase. Determination of the rate of the dose of gamma-radiation was conducted by ethylene and ferrosulphate dosimeters, and results of the measurements were agreed within the limits of 12-15%. The concentration of gas products of the radiation-thermal decomposition of mixtures of water with n-heptane was determined chromatographically on the devices Gasokhrom-3101 and Svet-102. In all cases kinetics of the process has been investigated, and the generation rate and the radiation-chemical yields of products were determined on initial segments of kinetic curves.

RESULTS AND DISCUSSIONS:

The radiation-chemical processes of decomposition of water, containing micro-impurity of n-heptane is studied in the temperature range of 100-400°C and absorbed dose of D = 1.8-10.8 kGy. The kinetics of gas production at the components ratio $[C_7H_{16}]/[H_2O] = 1.4 \ 10^{-3}$ and at the temperature of 400°C is shown on Figure 1.

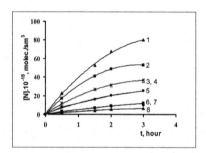

Figure 1. Kinetics of formation H_2(6), CH_4(2), C_2H_4(1), C_2H_6(3), ΣC_3(4), ΣC_4(5), ΣC_5(7), ΣC_6(8) at the radiation-thermal decomposition of C_7H_{16} in the water medium; $[C_7H_{16}]/[H_2O]=0.02$, T=400 °C, I=3.6 kGy/h.

It is clear, that at absorbed doses more than 7 kGy (for irradiation time 2 hours) have been observed tendency to the saturation. It is connected with decreasing of concentration of n-heptane in the mixture and decreasing of the rate of sensibilization of its decomposition by the water vapor. The total yield of gases in linear area of kinetic curves makes G = 379 molec/100 eV, that corresponds to the yield of the heptane's decomposition of 122 molec/100eV.

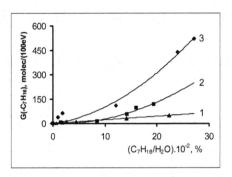

Figure 2. Influence of ratio of concentration $[C_7H_{16}]/[H_2O]$ to the radiation-chemical yield of decomposition of C_7H_{16} in the water medium; 1-100 ^{0}C, 2-300^{0}C, 3-400^{0}C.

The dependence of the radiation-chemical yield of heptane's decomposition on its initial concentration in the mixture with water at the different temperatures is shown on Figure 2.

In all range of studied temperatures the radiation-chemical yield of heptane's decomposition decreases with decreasing of concentration of n-heptane. This fact approves the above-stated explanation of the kinetics of gas generation at T = 400°C.

The temperature dependence of production and decomposition of C_7H_{16} within 100-400°C is depicted on Figure 3.

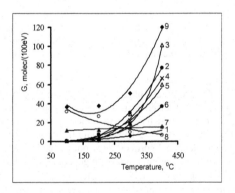

Figure 3. The temperature dependence of the radiation-chemical yields of gases: $H_2(1)$, $CH_4(2)$, $C_2H_4(3)$, $C_2H_6(4)$, $\Sigma C_3(5)$, $\Sigma C_4(6)$, $\Sigma C_5(7)$, $\Sigma C_6(8)$, $G_{(-)}(9)$; $[C_7H_{16}]/[H_2O]=14,2\cdot10^{-2}$%; J=3,5kGy/h.

As shown, the yields of all gases rise with temperature and their activation energy are: E (H_2) = 34.6, E (CH_4) = 34.6, E (C_2H_4) = 51.8, E (C_2H_6) = 38.5, E (C_3H_8) = 40.9, E (C_4H_{10}) = 44.3 kJ/mole. The temperature dependence of the rate of heptane's decomposition (Curve 7) is also studied. Thus the values of activation energy is equal E =11.3 kJ/mole.

The temperature variation of the radiolysis significantly influences to the ratio of yields of gas and liquid products (Table 1).

Table 1. Influence of temperature on formation of gas and liquid products.

T, °C	G_{gas}	G_{liq}	G_{gas}/G_{liq}
100	3.0	43.2	0.07
200	20.2	38.7	0.52
300	125.3	24.1	5.2
400	356.5	22.5	15.8

From the Table 1, it is clear, that with increasing the temperature from 100 up to 400°C the ratio G_{gas}/G_{liq} changes from 3 up to 15. It evident that at radiation- thermal cleaning of water of hydrocarbon products are separated to the gas phase, that is advantage of this process. At that effective activation energy of gas production is 11.3 kJ/mol. At the temperature higher than 300°C hydrocarbon component decomposes with negligible rate even at the thermal process. In that case the ratio of rates of thermal and radiation-thermal processes significantly depends on temperature (Table 2).

Table 2. The influence of temperature on ratio of radiation-thermal and thermal reactions rates.

T, °C	300	350	375	400
W_{tr}/W_t	101,2	55.2	26.1	1.9

Earlier we, on an example of pentadecane as a hydrocarbon model [5], have explicitly studied the relationships between that rate on temperature and dose rate, and have found optimum conditions for observation high sub-thermal effect in the radiation-thermal processes.

The observation of high values of radiation-chemical yields counted for a total absorbed dose by the system C_7H_{16}-H_2O, testifies the proceeding of sensitized radiolysis of C_7H_{16}. It is impossible to judge the precise mechanism of transfer of radiation energy from water to hydrocarbon now. Nevertheless it is possible to ascertain the fact that these processes flow through exited and ionized conditions of molecules of water. At the radiation-thermal decomposition of water the radiation-chemical yield of molecular hydrogen does not exceed 7-8 molec/100eV [6]. At the temperatures is higher than 300°C an atomic hydrogen generated at radiation decay of molecules of water (1. H_2O ▫ $H + OH$), reacts with a molecule of water and produces H_2 (2. $H + H_2O$ ▫ $H_2 + OH$). However at the presence of hydrocarbon there can be more easily way its consumption in the reaction with C_7H_{16} (3. $H + C_7H_{16}$ ▫ $H_2 + C_7H_{15}$). At presence of hydrocarbonic radicals at the high temperatures the chain regime of decomposition of hydrocarbon to the gas products can be originated. Thus it is possible to achieve high cleaning of water from oil. However, at very low concentrations of hydrocarbon (<10^{-2} %), the rate of a sensitization can decrease, and in these conditions the self-radiolysis of macro-components prevails. For example, at ratio concentrations of components [C_7H_{16}]/[H_2O] < 10^{-4} in a competition of reactions (1) and (2) can prevail (3). In these conditions the yield of products of a radiolysis is determined by decomposition of more radiation-stable component of H_2O. The competition of the ionic reactions should be studied separately in that case.

CONCLUSION:

At the temperatures higher than 300°C the radiation-thermal decompositions of hydrocarbon micro-impurities in water into gas products occurs according a chain mechanism and the radiation-chemical yield of the decomposition exceeds 100 molec/100eV. This method can be used for purification of sewages from oil contaminations.

REFERENCES:

1. Diederichen Christian. (2002). Reinigung von Prozessabwassern mittels Ultraschall. Erdol-Erdgas-Kohle 118 (10). 473-474
2. Nishigaki Kazu, Takeda Minoru, Tomomori Naotoka, Iwata Akira. Teion Kodaku (2002).J. Gryag.Eng. 37 (7). 43-49
3. Pikaev A.K. (2001). New Environmental Application of Radiation Technology. J. Khimiya vysokikh Energiy 35 (3). 175-187
4. Pikaev A.K. (2002). The Contribution of Radiation Technology to Environmental Protection. J. Khimiya vysokikh Energiy 36 (3). 163-175.
5. Mustafaev I., Gulieva, N. (1999). Radiation-thermal Transformation of Pentadecane. J. Khimiya vysokikh Energiy 33 (5). 354-359
6. Dzantiev B. G., Ermakov A.N., Jitomirskiy B.M., Popov V.N. (1979). Termoradiation Decomposition of Water Vapour. Atomnaya Energiya 46 (5). 359-361.

RADIATION-THERMAL REFINING OF ORGANIC PARTS OF OIL-BITUMINOUS ROCKS

I. MUSTAFAEV, L. JABBAROVA, N. GULIYEVA, K. YAGUBOV
Institute of Radiation Problems, Azerbaijan National Academy of Sciences,
31a H. Javid ave, AZ1143 Baku , Azerbaijan,

Corresponding author: imustafaev@iatp.az

ABSTRACT:

Oil-bituminous rocks (OBR) are one of the alternative sources of hydrocarbon raw materials and its world stocks in counted to hydrocarbon part are 250-300 Mt. Rational use of OBR as energy source and raw materials for chemical synthesis requires beforehand extraction of organic part from OBR. The hydrogen and hydrogen-containing gases obtained from OBR by using radiation and heat components of nuclear energy is interesting for solution of problems of nuclear – hydrogen energetic and effective use of solid fuel.

Keywords: Oil-bituminous rocks, liquid products, fractions, radiation and gases

INTRODUCTION:

Oil-bituminous rocks (OBR) are one of the alternative sources of hydrocarbon raw materials and its world stocks in counted to hydrocarbon part are 250-300 Mt [1]. Rational use of OBR as energy source and raw materials for chemical synthesis requires beforehand extraction of organic part from OBR. The obtaining of hydrogen and hydrogen-containing gases from OBR by using of radiation and heat component of nuclear energy is interesting for solution of problems of nuclear – hydrogen energetic and effective use of solid fuel. Joint impact of temperature and $\square$ - radiation (thermoradiation action) on process of formation fuel gases from shale and OBR has not been investigated. Existing data [2-4] refer to investigations of $\square$ - radiation effect on paramagnetism of Baltic shale "kerogen-70" [2], gases formation from Michigan's and Colorado's bituminous shale [3] and oil-bituminous rock Kirmaku deposit of Azerbaijan [4] at room temperature. In present work results of kinetic studies of hydrogen, carbon oxide and methane at thermoradiation conversion of OBR and OBR+H_2O systems are presented.

METHODS:

OBR samples from deposit at Kirmaku region of Azerbaijan has been exposed to thermoradiation conversion. Specific weight is 1.50-1.56 g/cm^3, oil contents in rock – 8.2-9.4 mass %, water saturation – 3.14%, clay contents in rock – on average 25.7%, H/C contents in organic mass – 0.14. OBR samples 0.1-1 g before irradiation are subjected to vacuum drying at 80°C during 0.5 h. Appointed quantity of water steam is given into ampoule with OBR from vacuum installation and is soldiered. Irradiation of ampoule has been carried out in $\square$ - radiation

M.K. Zaidi and I. Mustafaev (eds.), Radiation Safety Problems in the Caspian Region, 141-146.
© 2004 *Kluwer Academic Publishers. Printed in the Netherlands.*

source of Co^{60}, by dose power 40 kGy/h. Gaseous products has been analyzed by gas chromatography in "Gasochrom – 3101" apparatus.

EXPERIMENTAL RESULTS:

The kinetics of accumulation of fuel gases at thermoradiation conversion of OBR and OBR + H_2O systems has been investigated in temperature interval 20-500°C. The kinetics of accumulation of H_2 (1,2), CO (3,4) and CH_4 (5,6) at thermoradiation conversion of OBR (1,2,5) and OBR +0.1 MPa water steam system (2,4,6) at 400°C is presented in Fig. 1.

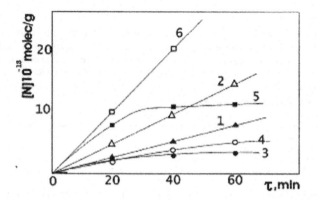

Fig 1. The kinetics formation of H_2 (1,2), CO (3,4) and CH_4 (5,6) at thermoradiation conversion of OBR (1,3,5) and OBR + H_2O (2,4,6) systems; J = 40 kGy/h, t = 400°C, P(H_2O) = 0.1 Mpa

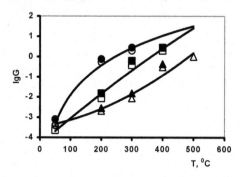

Fig 2. The temperature dependence of radiation – chemical yield of H_2(1), CO(2) and CH_4(3) at thermoradiation conversion OBR (Δ,O,□) and OBR + H_2O (▲,●,■);J = 40 kGy/h, P(H_2O) = 0.1 MPa.

Radiation-chemical yield of gases on initial parts of kinetics curves (number of produced molecules per 100 eV of absorbed energy, G molecule/100eV) are equal to G_1=2.7, G_2=5.3, G_3=1.1, G_4=1.6, G_5 = 4.1, G_6 = 12.5. The temperature dependence of rate of gas formation is presented in Fig.2.

Increasing of temperature from 20°C to 500°C leads to increasing of radiation – chemical yields of gases from $G(H_2)$ = 0.054, $G(CO)$ = 0.003, $G(CH_4)$ = 0.004 to $G(H_2)$ = 52.1, $G(CO)$ = 3.9, $G(CH_4)$ = 62.5. In this case energy activation of gases formation are: $E(H_2)$ = 118.2, $E(CO)$ = 32.5, $E(CH_4)$ =7 6.4 kJ/mole.

Yields of gases are not changed by addition of water steam (P_{H2O} = 0.1 MPa) to OBR in temperature interval 20 – 300°C, however at increasing temperature to 500°C yields of hydrogen and CH_4 increase up to $G(H_2)$ = 95.1, $G(H_4)$ = 142.6. In this case $G(CO)$ does not change significantly.

In the dependence of radiation – chemical yields of fuel gases from OBR + H_2O (at temperature 400°C) on water steam pressure P = 0-0.025 MPa (Fig.3) is observed monotonous increasing of yield of H_2 and CH_4 to $G(H_2)$ = 7.1, $G(CH_4)$ = 16.2. Further growth of $P(H_2O)$ to 0.2 MPa do not change $G(H_2)$ and $G(CH_4)$.

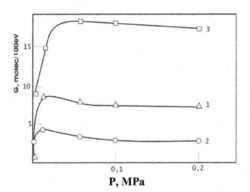

Fig 3. The dependence of radiation-chemical yields of H_2(1); CO (2) and CH_4 (3) from water steam pressure at thermoradiation conversion OBR + H_2O systems; t = 400°C, J = 40 kGy/h

Formation of H_2, CO and CH_4 at radiation decomposition of OBR is related with radiation decay of organic part of OBR(1) and recombination (2) reactions of primary radicals:

OBR $\xrightarrow{\sim\!\!\sim}$ H, CH_3, CO   (1)

H+H+M ◻ H_2+M   (2)

CH_3+H ◻ CH_4

Radiation chemical yields of gases at room temperature don't exceed value $\square G^g$ = $G(H_2)$ + $G(CO)$ + $G(CH_4)$ = 0.06 owing to high radiation stability of aromatic part of organic mass of OBR. Such value of gas yields are observed early at the $\square$-radiolysis of coal and oil-residues [4]. Yields of gases are rising by proceeding of secondary process (3) and decomposition of weak combined oxygen containing groups:

$$H+OBR \square \quad H_2 + R^{'} \text{ and } CH_3+OBR \square \quad CH_4+R^{'} \qquad (3)$$

Significant increasing of $G(H_2)$ and $G(CH_4)$ in presence of water steam (Fig. 3), apparently, is related with formation of active particles (H, OH, H_2O^*) occurring at the radiolysis of water:

$$H_2O \longrightarrow\!\!\wedge\!\!\wedge\!\!\blacktriangleright H, OH, H_2O \quad (4)$$

Chain processes of formation of fuel gases proceed at thermoradiation conversion of OBR and OBR+H_2O systems in the temperature range T $\square$ 400^{0}C and at 500^{0}C chain length reaches to $\square$ $\square$ G_g/G_R $\square$ 10^2. Fuel gases are formed also by direct thermal decomposition of OBR at temperatures T > 250°C. However, as seen from fig 4, rate of thermoradiation (W_{TR}) process is higher than rate of thermal (W_T), or else W_T/ W_{TR} < 1 for all investigated temperature interval (T $\leq$ 500° C). Limiting stage of initiation (1) is overcome by energy of ionization irradiation. This leads to conversation of the inequality W_T/ W_{TR} in all investigated temperature intervals, however with increasing T value of this correlation increases (see Fig.4)

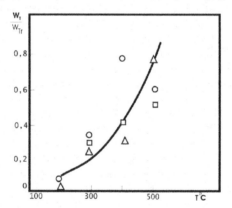

Fig 4. The dependence of W_T/W_{TR} from temperature at thermal and thermoradiation conversion of OBR; Δ –H_2, O –CO, $\square$ – CH_4, J = 40 kGy/h.

DISCUSSION:

In thermal and radiation initiation of chain processes conditions, decomposition of organic part of OBR on the limits suggested scheme of reaction (1-3) rate of process in some assumptions is determined by expression

$$W_{TR} = \frac{k_3}{k_2^{1/2}} \sqrt{v_t + v_r} \qquad (5)$$

$v_t = k^0 e^{-E_3/R} [RH]$
$v_r = G_R J \; 10^{-2}$

k_i – rate constant of reaction described by equation (i = 2,3)

$v_t = k^0 e^{-E_3/RT} [RH]$ – rate of thermal decomposition

$v_r = G_R J \; 10^{-2}$ – rate of radiation-chemical decomposition

k^0 – pre-exponential coefficient of rate constant

E_3 – activation energy of reaction (3)

E_0 – activation energy of thermal decomposition of OBR

G_R – radiation-chemical yield of radicals

J – dose rate of the radiation

[RH] – concentration of steam of organic compounds

Submitting these values to equation (5) we can obtain expression for W_{TR}:

$$W_{TR} = \frac{k_3^0 e^{-E_3/TR}}{k_2^{1/2}} \sqrt{10^{13} e^{-E_0/RT}[RH] + G_R J \cdot 10^{-2}} \qquad (6)$$

Consequently

$$\frac{W_T}{W_{TR}} = \sqrt{\frac{1}{1 + \dfrac{G_R J \cdot 10^{-2}}{10^{13} e^{-E_0/RT}}}} \qquad (7)$$

These expressions satisfactorily describe the dependence of thermoradiation gases formation rates and ratio W_T/W_{TR} on temperature, dose rate, and concentration of steam of organic part of OBR. As seen from (7), at fixed value of J with increasing temperature W_T/W_{TR} value is rising. The optimal intervals of temperatures and dose rates for radiation-thermal processes have been defined in [5] on example of pentadecane decomposition. It's necessary to note, that in strong sources of ionization (electron accelerator, nuclear reactor) significant high dose rate (J) may be reached as relatively and as absolutely values of rate of thermoradiation process of oil-bituminous rock conversion may satisfy to requirements of technology.

CONCLUSION:

The radiation - thermal effect is one of effective method for refining of oil-bituminous rocks to fuel gas and organic products. The contribution of radiation effect at the radiation-thermal processes depends on the ratio of temperature and dose rate and in certain conditions it can be more than 90 %.

REFERENCES:

1. Oil-bituminous rocks. Perspectives of utilization. The materials of all union conference on complex conversion and utilization oil-bituminous rocks. Nauka. Kazachskaya SSR, Alma-Ata, p.3, 1982.
2. Yakovlev V.I., Chinman R.E., Shpilfogel P.V., Sendjurov M.V. in: Investigation in chemical area and technology of products conversion of fuel minerals. LTI by name Lensovet, v.1, p.12, 1974.
3. D'Anjou D., Litman R. Radiochemical and Radioanalytical letters, 50, No1, p.37-44, 1983
4. Mustafaev I.I., Yakubov K.M., Gadjiyev Kh.M., Guliyeva N.K. Radioanal. Nucl. Chem. Letters 94 /I/1-8, 1985.
5. Mustafaev I. I. , Guliyeva N. K. Khimiya Vysokikh Energiy, v 33, No 5, p.354-359, 1999.

RADIO-ECOLOGICAL STATE OF
ABSHERON OIL AND GAS EXTRACTING FIELDS

R.N. MEHDIYEVA, H.M. MAHMUDOV, M.F. GAFFAROV
Institute of Radiation Problems, National Academy of Sciences
159 Azadlig Avenue, AZ1106, Baku, AZERBAIJAN

Corresponding author: hokman@rambler.ru

ABSTRACT:

Oil and gas have been extracted for some time in Absheron peninsula. Oil-contaminated soils and the polluted ponds, created close to such spots, have the worst effect for the weather, fauna, flora and the health of the inhabitants. The radiation background measurements conducted on the type and quantity of radionuclides in the fields and to decontaminate them. It was found that the radiation background in the oil polluted water pools and soils in some areas of oil extracting fields are much more than the natural levels.

Keywords: radiation background, exposure dose rate and radioactivity.

INTRODUCTION:

Traditionally oil and gas have been extracted for a long time in Absheron peninsula. Around the oil wells there are oil-contaminated soils and the polluted ponds, it stands, that they have the worst effect for the weather, fauna, and flora of Absheron and the health of the inhabitants living there. The pollution of biogeocenosis by radioactive matters causes new mutagen and evaluation factors having influence to the living organisms' coexistence within the population for a long time to come. The influence of radionuclides in the natural biogeocenosis to the organisms' coexistence and population has given rise to the new field of radioecology, a sub discipline of general-ecology. With the influence of environmental protection ideals, radioecology is important for the study of the influence of radiation to the biosphere's components, the regulation of radionuclides migration, bioaccumulation mechanism, and determination of the technological increase of natural radionuclides and so on. This is one of the most important directions and has the greatest increasing perspective.

As a result of biosphere pollution from radioactivity the studying and safeguarding of environment are the most actual and necessary problem. Highly intensifying the matter of the ecological crisis is the increasing of radioactive wastes year after year. The intensive development of the atomic industry, making of powerful electron accelerators, implantators, plasmatrones, radiation manufacturing plants and production of radioactive elements regularly raise the radiation background by spreading constant anthropogengenic and technologically enhanced natural radionuclides. This makes for the greatest excitement and concern.

M.K. Zaidi and I. Mustafaev (eds.), Radiation Safety Problems in the Caspian Region, 147-150.
© 2004 *Kluwer Academic Publishers. Printed in the Netherlands.*

The migration and regulation of these natural and unnatural radionuclides in the biosphere depends on numerous factors. That is clear, the radionuclides dropped to the environment, very easily migrate in the line of "weather-soil-plant-animal-human-being" and enter into the biological circulation. The main ring of this action is "soil-plant". It mainly determines the general migration intensity in biosphere. In this case the exact factor being able to influence to the speed of the action is the physio-chemical properties such as; its mechanical, mineral content, saltness, separate radionuclides of plants, ability dealing with selecting and gathering of chemical elements and so on. There is too little information about the migration speed depending on mineralization, dampness and pH of the nuclide elements in various soils being migrated. It stands such information is so necessary, because to give prognosis about the further tale of radioisotopes without this is impossible. Probably as a result of such investigations, not only to radioecology and its close fields, but also the theoretical foundations of this science will be given a possible prognosis linking to the ecological results of "nuclear energy" complex industry's development will be improved.

METHOD:

Radiometric measurements have been done with scintillation detectors, which determine CPS of the 88H plant in open-air reaches. Due to the bioaccumulation of radionuclides in this plant it is used for determination of radioactivity in soil and mountain layers indirectly. By measuring the gamma ray intensity in the area of 3881 m^2 / (s^{-1} m q^{-1}) the gamma intensity ranges between 10-3×10^4 s^{-1}. The special background of the plant isn't more than 10 s^{-1}. This strategy is considered suitable and valid for the open-air reaches.

RESULTS AND DISCUSSION:

The studying of the radio-ecological state of the oil-contaminated soils and the ponds polluted is the most actual problem at present. The research of the radiation background made by the radioactive nuclides accumulated for years in the local areas of oil extraction and its initial refining. At the same time, the studying of these types, and quantity of radionuclides comprising this background are very important for the usage of these areas. This is why research is being done to measure the radiation background in the necessary fields and to decontaminate it.

The oil-contaminated soils investigated were the major areas where the oil extracting, transporting and the initial refining have been realized. The expanse of these areas is too large and surrounds Qaradag, Sahil, and Sabunchu, Azizbeyov regions. In general, due to the conclusions based on all measuring and references, the middle natural radiation background is equivalent to the exposure 7-8 µR/hour. In spite of this, there are some fields, used for many years, where the radiation made by the radioactive nuclides gathered is much more than the natural background. The results of the research show that mainly in the areas where oil is extracted, the radiation background is different in comparison with the normal one, sometimes the radioactivity of these fields is about 500-1000 µR/hour. The area of Bibi Heybet Oil-Gas Extracting Enterprise is characterized with this feature. We dare say, this area has been entirely polluted by the sediments of (ground water) the water of the oil wells.

The radiospectrometric analysis of the soil samples taken from the northborder of the location having near 1000 µR/hour radiation has revealed the existence of the radium isotope in the oil-contaminated locations.

In the area of Qarachuxur Oil-Gas Extracting Enterprise, the fields polluted having near 1000 µR/hour radioactivity have been revealed. In the settlement of the workers, such pollution made the yellowish, soft, friable soils. For the shape, these appeared with in the pipe while cleaning the pipes up, were demolished on the ground.

Here in the area of oil field there are sites having 700, 600, 300, 150-200 µR/hour radioactivity. These are the areas being characterized with high radiation background (30-25µR/hour) in the fields of the oil-gas extracting enterprise named after Azizbeyov.

Due to the measuring in Pirallahi, Balakhanineft, Binegedineft and Buzovnaneft, the enterprise found that these are not the areas with high radioactivity, in these locations the radioactivity is about natural one. Not only in the fields of oil and gas extracting enterprises but also in some settlements, there are locations characterized with high radiation background. In the area of oil plant in Hervsani, the zones with 10µR/hour radioactivity were revealed.

There're the abnormal locations having near 150µR/hour radiation background, around the oil diluting mediums in the settlement. In the area next to the iodine plant in Ramana, radioactive background is about 100 µR/hour.

The areas of the mechanical engineering plant, its all sections, and canteen, lab being built have been leant in details.

In this period, in everywhere the radiation backgrounds 8-10 µR/hour but in the area of the lab that will be built with the red brick, the radiation background is 20 µR/hour.

In the Yeni Surakhani settlement local area, 40mR/hour radioactivity levels have been revealed. As the result of the latest investigation, the distribution of gamma rays' exposure dose rate gathered in the oil and gas extracting enterprises' local areas in Absheron peninsula have been shown in the Figure 1.

CONCLUSIONS:

◻ In Absheron peninsula the total radiation background is 8-10 µR/hour.

◻ Some local areas having higher radiation background than normal one must be decontaminated.

◻ In the area of oil and gas extracting fields, the research should be conducted. The radiation background be measured and recorded carefully for any changes.

150

❑ The lands having high radiation background have to be decontaminated.
❑ The radiation background should be studied while leasing lands or estimating soils.

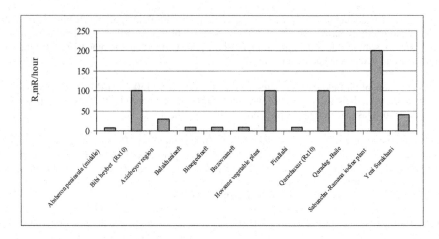

Figure 1. The distribution of gamma rays' exposure dose rate gathered in the oil and gas extracting enterprises' local areas in Absheron peninsula.

REFERENCES:

1.   Polyakov V.A., et.al. (2001). Komplexniy podxod k izuchrniyu radioekologicheskoy obctanovki na territoriyax neftepromislov 1. 26-36
2.   Soloduxin V.P., Kazachevskiy, Reznikov S.V. (2000). Izmereniya urovney radiaktivnosti pri dobiche, podgotovke I transportirovke gazoneftyanogo sirya. Apparatura I novosti radiasionnix izmereniy (ANRI) 3. 10-14.
3.   Aliyev L.A., Khudaverdiyeva. (2003). Radiation Monitoring of the Environment: measurements of the background and aerosols in territory of Azerbaijan Republic. Int.Congress on the Protection of the Environment from the effects of Ionizing Radiation, Stockholm, Sweden.

INFLUENCE OF TECHNOLOGICAL CYCLES OF NATURAL GAS TREATMENT ON RADIOACTIVE RADON CONTENT IN ITS COMPOSITION

A.A. GARIBOV, G.F. MIRALAMOV*, R.CH. MAMEDOV*, G.Z. VELIBEKOVA.
Institute of Radiation Problems, NAS of Azerbaijan
*Azerbaijan Gas Processing Plant
H.Javid 31a Avenue, AZ-1143, Baku, Azerbaijan.

Corresponding author: hokman@rambler.ru

ABSTRACT:

Radioactive gas radon is found in the composition of natural gas. Radon is formed as a result of radioactive decay of natural radionuclides and it goes into the natural gas. Despite the fact that certain part of radon volatilizes or decays in the process of natural gas production, transportation and processing even then a great quantity of radioactive substances come to damage the environment and effect human life. The investigation of natural gas during its processing are of great interest.

Keywords: Radon, radionuclides, natural gas, oil and kerosene oil.

INTRODUCTION:

The gas radon is in the composition of natural gas. Radon is formed as a result of radioactive decay of natural radionuclides and it goes into the natural gas. Despite the fact that certain part of radon volatilizes or decays in the processes of natural gas production, transportation and processing due to the natural gas the great quantity of radioactive substances come to harm life. The investigations of regularities of the radon content changes in the composition of natural gas during its processing. The regularities of radon content changes in the composition of natural gas depending upon deposits, depth production and at different stages of its processing have been firstly studied.

The absorption method is used at gas-processing plants for the extraction of heavy hydrocarbon fractions from the natural gas. Kerosene oil fractions are used as a sorbent. Relatively heavy hydrocarbons are isolated from sorbent by the desorption method and later they are isolated in separate fractions. It is established that the values of volume activity for radon are changed in wide range from 20 to 1178 Bq/m^3.

The following conclusions have been done on the basis of obtained results: the content of radon in the composition of natural gas coming to the process is changed in wide range. It's contents of radioactive isotopes are very different at different gas fields.

M.K. Zaidi and I. Mustafaev (eds.), Radiation Safety Problems in the Caspian Region, 151-155.
© 2004 *Kluwer Academic Publishers. Printed in the Netherlands.*

Kerosene oil fraction of the oil is used as a sorbent during gas processing and it has absorption activity by radon. The content of radon in consumption gas decreases up to $25Bg/m^3$ as a result of absorption processes of treatment of natural gases from relatively heavy hydrocarbons.

It's known that natural gas contents radioactive gas radon [1]. Radon penetrates in natural gas from contacted with it grounds, where it is formed as a result of radioactive decay of natural radionuclides. Despite that certain part of radon is evaporated or decomposed during the process of production, transportation and treatment of gases, great volume of radioactive substance [2-5] comes in life conditions due to natural gas. That's why investigation of change regularity of radon content in natural gas during the process of its treatment is of great interest.

At present time in Azerbaijan large gas accumulations are investigated and their exploitation has already begun. Planned gas production pace of development in Azerbaijan as well as all over the world creates necessity of radon content investigation in natural gas composition depending upon field, production depth on different stages of its treatment. Regularities of radon content changes in composition of natural gas and product of its treatment on gas-processing plant are investigated for the first time   in the given work.

METHODS

Activity concentration (AC) of radon (^{222}Rn) in composition of natural gas is determined by radiometer PPA-01M. ^{222}Rn AC measurement is based on electrostatic precipitation of charged ions of ^{218}Po(RaA) from controlled gas medium on the surface of semiconductor detector. At this ^{222}Rn AC is determined according to the quantity of registered alphas during RaA decay. In complex with sampler radiometer allows to determine radon AC in liquid and solid systems. Measurement range of ^{222}Rn is from 20 to $2 \cdot 10^4$ Bq/m^3. Radiometer sensitivity is $(1.61 \pm 0.32) \cdot 10^{-4}$ c$^{-1} \cdot$Bq$^{-1} \cdot$m^3. Inaccuracy of measurements is $\pm 30\%$.

Radioactivity of liquid and solid samples is studied also by α-specters on АИ-1024 analyzer with meter on the basis of NaI(Tl). Investigated samples are placed in special constructed lodge with defense from external radiation. Information comes from analyzer АИ-1024 via synchronizer unit (SU-2) to computer. Activity of radioactive isotopes U-238 (Ra^{226}), Th-232 (^{228}Ra, ^{224}Ra) and ^{40}K. Together with abovementioned methods changes of equivalent dose of α-radiation and fluencies of α-particles with the help of radiometer MKC-01P with detector БДКА-01P.

RESULTS AND DISCUSSION:

On gas-processing plant, the gas comes to natural gas separation set from comparatively sinking fractions. Absorption method is used for separation of comparatively sinking hydrocarbons fractions from natural gas. Oil fractions are usually used as sorbents. Comparatively hydrocarbons are separated from sorbent by desorption method and are later divided to separate fractions. The objective and volume of investigation of radon content change regularity in composition of natural gas and in its treatment products are determined by technological peculiarities of the plant. Initially radon content in composition of feed stock – natural gas had been investigated. Results of investigations are shown in Table1. As it is seen from the table radon AC meanings in feedstock are changed in wide interval 20-1178 Bq/m^3. Observed variations in radioactivity meanings may be connected with different radon content in natural gas in different fields.

Gamma-spectrometry analysis of soil from the depth of gas fields testifies the presence of U-238 (Ra-226), Th-232 (Ra-224) in them.

Table 1. Radioactive radon content in composition of natural gas, coming to gas-processing plant

№	Analysis date	Activity concentration of natural gas, Bq/m^3
1	25.03.2002	1178
2	25.03.2002	1110
3	28.03.2002	345
4	02.04.2002	37
5	29.04.2002	76
6	25.04.2002	85
7	12.04.2002	23
8	19.04.2002	≤20
9	01.05.2002	76

Radon isotopes are formed as result of three rows of natural radioactive decay:

$$U\text{-}238 - {}^{226}Ra_{88} \xrightarrow{\alpha} {}^{224}Rn_{86} \text{ (radon)}; T_{1/2}=3,64 \text{ days}$$

$$Th\text{-}232 - {}^{224}Ra_{88} \xrightarrow{\alpha} {}^{220}Rn_{86} \text{ (thoron)}; T_{1/2}=54,5 \text{ seconds}$$

$$U\text{-}235 - {}^{223}Ra_{88} \xrightarrow{\alpha} {}^{219}Rn_{86} \text{ (actinon)}; T_{1/2}=3,92 \text{ seconds}$$

Lifetime of thoron and actinon is little and so they are evidently decayed during production and transportation. That's why observed radioactivity of natural gas at the entrance to technological cycle is conditioned by the presence of $^{224}Rn_{86}$.

The investigation has been carried out on three technological units of the same type and five sets. Analysis of radon content at the entrance and exit of technological cycle of gas sorption had been carried out on each set. Analysis at the exit of sorption settings was carried out every hour during 8 hours. Table 2 shows the results of analyses at conditional time $\tau_1 < \tau_2 < \tau_3$, where $\tau_1 \leq 60$ min, $\tau_2 \square 90$ min, $\tau_3 = 240$ min. As it is seen radon content in natural gas composition increases with increase of time.

Let's suppose that in experimental conditions absorption activity of kerosene gas oil fraction on radon a_m and active centers are equally distributed on fraction volume.

$$Rn + a_m \square \quad a_m - Rn \text{ (absorption state of radon)}$$

$$(1)$$

Interaction between absorbed radon atoms and influence of a_m – Rn – complexes on free states of Rn can be neglected.

Table 2. Change of radon activity concentration (RAC) in composition of natural gas as a result of absorption processes.

Technological unit	Setting	Samples for analyses	Date of sampling	RAC in composition of natural gas	RAC in composition of gas emanating from setting
1	2	Natural gas, emanating from setting	01.05.02	76	(τ_1) - ≤ 20 (τ_2) - 74 (τ_3) -134
2	3	Natural gas, emanating from setting	01.05.02	76	(τ_1) - ≤ 20 (τ_2) - 38 (τ_3) -71
2	4	Natural gas, emanating from setting	01.05.02	39	(τ_1) - ≤ 20 (τ_2) - ≤ 20 (τ_3) - 45
3	5	Natural gas, emanating from setting	01.05.02	41	(τ_1) - ≤ 20 (τ_2) - 121 (τ_3) - 145

Sorption processes of radon (1) may be described by Langmuir equation:

$$a_{Rn} = \frac{a_m k a_2}{1 + k a_2}$$

Where a_{Rn} is current concentration of occupied active states for radon sorption; a_2 – radon content in composition of natural gas; k – equilibrium constant of adsorption and desorption (1).
At initial changes $k a_2 \ll 1$ and $a_{Rn} = a_m k a_2$, i.e. radon sorption linearly depends on radon concentration in natural gas. When $k a_2 \gg 1$ radon sorption doesn't depend on its content in composition of natural gas and sorbent saturation with radon takes place. That's why after some definite time radon sorption ceases. As in kerosene gas oil fraction radon holding intensity is within the range of Vander-Val interaction forces, after saturation gas flow may take retained radon atoms. That's why radon content in composition of emanated gases at a lot of time sometimes exceeds radon content in incoming gas. Time of one absorbent circulation cycle in absorbers is equal to 1 hour. As it is seen from table 2 during this time effective separation of natural gas from radon via sorption on kerosene gas oil fraction takes place. If we'll consider kerosene gas oil fraction and natural gas as two component system it's possible to define radon dissolubility (◻) in absorbent in conditions of technological cycles [6].

$$◻ = \frac{E_{ma◻} l\, V_{a◻}}{E_{m_2} l\, V_2}$$

Where $\square$ is radon dissolubility that corresponds to the ration of its concentration in absorbent to the concentration in air; $E_{m_{a\square}}$ - emanation quantity in absorbent; E_{m_2} - emanation quantity in gas; $V_{a\square}$ and V_2 – volumes of absorbent and gas phase.

In technological processes $V_2=10^5$ m^3/h, $V_{a\square}=100$ m^3/h, $E_{m_2}=76$ Bq/m^3, $V_{a\square} = 22,00\square 0^3$ Bq/m^3, basing upon this $\square=0,29$ is determined.

Observed $\square$ meanings differ greatly from the meanings of radon dissolubility in kerosene, observed in [6]. Radon dissolubility decreases in flowing system and conditions of absorption technology.

CONCLUSION:

$\square$ Radon concentration in composition of gas, incoming for processing changes within wide range 26 - 1180 Bq/m^3. To all appearance this is connected with different content of radioactive isotopes on different gas fields.

$\square$ Kerosene gas oil fraction that is used as absorbent in gas processing has absorbent activity according to radon.

$\square$ As a result of absorption processes of natural gas treating from comparatively heavy metals, radon concentration in consumed gas decreases up to <25 Bq/m^3.

REFERENCES:

1. Radiation, dose, effects, risk. Translation from English, Moscow, "Mir", 1990,p.28
2. Weers, A.M.. Van pace,1. Strand T., Lycebo I., Watkins S., Sterker T., Meijne E.I.M. Butter K.R. (1997) Current practice of dealing with natural radioactivity from oil and gas production in EU member states, EU 17621 EN, EU Directorate-General Environment, Nuclear Safety and Civil protection, European Commission, Luxemburg
3. Gessel T.F. (1975), Occupational radiation exposure due to 222-Rn in natural gas and natural gas products, Health Physics, Vol.29, p.681-687.
4. Otto G.H. (1989). A National Survey on Naturally Occurring Radioactive Materials (NORM) in Petroleum Production and gas facilities, The American Petroleum Institute, Dallas, Texas.
5. Jonkers G., Hartog F.A., Knaepen A.A., Lancee P.F.(1997). Characterization of NORM in oil and gas production (E&P) industry, Proc. Int. Symp. On radiological problems with Natural Radioactivity in the Non-Nuclear industry, Amsterdam 8-10 September 1997, KEMA, Arnhem 6. Reference-book on radiometry, M.Gosgeoltechizdat, 1957, p.74

MARINE GAMMA SURVEY OF SEABED OF CASPIAN SEA

Z.T. BAYRAMOV and A.M. GASANOV
Azerbaijan National Aerospace Agency
159 Azadlig Avenue, AZ 1106, Baku, Azerbaijan.

Corresponding author: bayramov_z@yahoo.com

ABSTRACT:

A new global project was developed to exploite transportation of hydrocarbons, restoration of Great Silk Way, exploitation of new nuclear reactors and radiation technology to organize a permanent radiation monitoring and control system for the environment. In particular, in the territory of Azerbaijan Republic including a sector of Caspian Sea, have observed irreversible processes that are the result of anthropogenic influences on environment.

Keywords: gamma survey, radiation map, and underwater gamma spectrometer.

INTRODUCTION:

The industrial waste, created by activities such as oil deposit prospecting and exploitation, work on buildings, lying of oil pipelines, is dumped in Caspian Sea. The waste get in by Volga river on its banks main nuclear industry enterprises of Russian Federation are situated, and a nuclear desalinized plant in Aktau city (Kazakhstan) contribute some. All these waste have increased the radiation background level of Caspian Sea. The Caspian Sea is the world's largest inland water body with a surface area of 37900 km^2, a drainage area of approximately 3.5 million km^2, and a volume of 78000 km^3. The Caspian Sea is divided into three basins with approximately equal surfaces. The northern basin has a maximum depth of 5-15 m, and contains 1% of the total water. The central basin has a maximum depth of 800 m and contains 22% of the total water. Southern basin which has a maximum depth of 1024 m and an average depth of 330 m, and contains 77% of the total water of the Caspian Sea. To explore the potential of isotope techniques for studying water balance and dynamics - two training courses were conducted by IAEA [1].

Unfortunately during these expeditions there is a highly important aspect of the problem of maintaining of ecological balance, which is covered insufficiently. This is composition of seabed's radiation situation map. This question gains the special important for exploitation of closed systems, which include also the Caspian Sea. The problem is that the Caspian Sea unlike to other water basins possessing rich deposits of hydrocarbon raw materials (e. g. the Mexican Gulf, the North Sea, the Persian Gulf), is a closed type bio-ecological system, where the rigid balance of hydrosphere, biosphere and especially of geological structure of seabed is existed. The violation of any of these components can lead to practically irreversible processes, and as a result, to ecological catastrophe. Realization of gamma survey of seabed makes it possible to determine initial background of seabed's gamma field at the beginning of well drilling work.

M.K. Zaidi and I. Mustafaev (eds.), Radiation Safety Problems in the Caspian Region, 157-160.

Further periodical monitoring of radiation situation will be based on results of initial condition of seabed's gamma-field.

METHOD:

By this purpose we developed an original device for determination of percentage content of major components of gamma field, namely, of natural radio nuclides U-238, Th-232 and K-40 in three working energy windows: 1.65-1.85, 2.5-2.8, and 1.35-1.55 MeV using Cs-137 calibration (0.662 MeV) – (Figure 1).

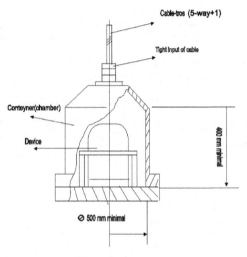

Figure 1.

Usually gamma- spectrometric analysis of contents of natural radioactive elements in situ at use of spectrometers with scintillation detectors (for example, NaI (Tl)) is spent in a hard part of a gamma-spectrum, where, there are intensive and good resolved photo peaks: 1.46 MeV from K-40, 1.76 MeV from Bi-214 and 2.62 MeV from Tl-208. Thresholds of analytical channels corresponding to these peaks get out so that at the maximal sensitivity of the analysis the error of definition of concentration of natural radioactive elements was as small as possible. Optimum values of thresholds depend on type of a spectrometer and conditions of measurement, however practically differ from 1.3-1.6 MeV ▢potassium▢, 1.6-2.0 MeV for ▢uranium▢ and 2.4-2.8 MeV for ▢thorium▢. Spectrum of natural gamma-radiation in area of energy E >1.2 MeV is simplified it is possible to present as the sum of spectra from three natural radioactive isotopes - sources of monochromatic gamma-quanta and a spectrum of background radiation [2]. If U and Th are in deposits in a status of radioactive balance with products of their disintegration between speeds of the bill of pulses in analytical channels and contents U, Th, K it is possible to write down connection in the following kind:

$$Q_{Th} = R_1 (N_{Th} - B_{Th}), Q_U = R_2 (N_U - S_1 \overline{N}_{Th} - B_U), Q_K = R_3 (N_K - S_2 \overline{N}_U - S_3 \overline{N}_{Th} - B_K)........ (1)$$

Q_{Th}, Q_U, Q_K - accordingly contents (equivalent) thorium, uranium and potassium in a surface layer of deposits of 0 - 45 cm; N_{Th}, N_U, N_K - rate of pulses in corresponding channels of a spectrometer; B_{Th}, B_U, B_K - background rate of pulses in corresponding analytical channels; S_1, S_2, S_3 - coefficients which are taking into account the contribution of scattering gamma-radiations; $\overline{N}_{Th} = N_{Th} - B_{Th}$ - ☐true☐ rate of pulses in ☐thorium☐ channel, caused only by gamma-radiation of thorium; $\overline{N}_U = N_U - S_{1Th} - \overline{N}B_U$ - ☐true☐ rate of pulses in ☐uranium☐ the channel, caused only radiation of uranium (is more exact, Bi-214); R_1, R_2, R_3 - coefficients of transition from ☐true☐ rate of pulses to contents of natural radio nuclides in deposits. Definition of coefficients in the formula (1) is connected with the big difficulties. Therefore gamma - spectrometer calibration are usually carried out on result of special measurements on models - sources of gamma-radiations or in natural conditions. The most exact values calibrate coefficients turn out in the latter case as thus all parameters of the detector and a condition of scale - shooting are completely taken into account. As a result of calculation on computer under specially made program the ratings of coefficients and the following formulas for calculation of equivalent contents of natural radio nuclides in sandy deposits (average humidity - 35 of %) by results of gamma- spectrometer have been received:

$Q_{Th} = 1.8 (N_{Th} - 0.5)$, $Q_U = 1.65(N_U - 0.75\overline{N}_{Th} - 0.4)$, $Q_K = 0.15(N_K - 1.2\overline{N}_U - 0.94\overline{N}_{Th} - 0.9)...(2)$

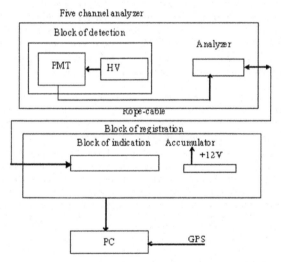

Figure 2.

These background rate of pulses for analytical channels appeared relatives on size to the measured rate of pulses at an arrangement of the detector of a spectrometer in a thickness of water on distance from a bottom more 3 m ($N_{bTh} = 0.38$, $N_{bU} = 0.52$, $N_{bK} = 1.0$ imp/s).

Underwater part designed from hermetic container. The five-channel gamma-spectrometer will consist of the block of registration and analyzer. The block of registration is intended for transformation of energy of gamma-quanta to an electric voltage's pulses. The block of registration consists of photomultiplier tube (PMT), non-organic NaI (Tl) scintillator and high voltage block. The pulse of an electric signal from an output of PMT acts on an input of analyzer. The analyzer divides entrance pulses on their amplitude in such a manner that on his output the pulses corresponding to energy of isotopes of potassium, uranium and radium were formed. These signals on a cable are transferred to the shipboard s block of registration. The block of registration serves for visualization of quantity of the pulses registered in energetic windows, corresponding to energies of isotopes potassium (K-40), uranium (U-238) and thorium (Th-232). Simultaneously ship computer accepts these signals and processes them. These measurements are conducted from a stationary position of ship. Time of measurement in each point makes not less than 30 minutes. Results of radionuclide measurements ranges of energies; 1.35-1.55 MeV (K-40), 1.65-1.85 MeV (U-238) and 2.5-2.8 MeV (Th-232) for each point of ship stop together with its geographical coordinates are brought in a database of PC. The definition of a site of a ship is carried out by Global Positioning System (GPS). On the basis of a database the radio ecological map of an initial condition of a sea bottom is under construction. In each point of a stop of a ship with the help of the special device selection of test of sediment is made. These tests demand more careful analysis, that is much greater time of measurement, and it is possible only in stationary conditions.

The information received from the low depths transfers to the scientific-research craft using a rope-cable and treat on the ship's computer. The reliability and possibility of long period of continuous work of proposed device allows to carry out mapping of area's radiation situation at sea conditions along predetermined route, the co-ordinates of which are determined by high-precision navigational satellite system DGPS. Data received from synchronous measurements of radiation background and geodesy co-ordinates were analyzed. The special software for processing make it possible to develop the map of radiation situation of pre-arranged scale for determined area and record the initial condition of environment.

CONCLUSION:

Gamma-mapping of seabed and sediments make it possible to decide two problems. The first is an ecological problem - the reason of an increase of seabed's natural radioactive background. The second is oil and gas deposits exploration. Researches have shown that natural radioactive background existing over the oil and gas deposits contours is approximately 3-5 times lower than that of other areas as the leakage of radionuclides to upper layers in oil, is comparatively low. This process causes the shortage of natural radionuclides in upper layers of seabed. The contours of oil and gas deposits situating in depth of seabed may be accurately determined using developed maps [3].

REFERENCES:

1. Froehlich K.et.al. (1999). Isotope studies in the Caspian Sea. IAEA. In The Science of the Total Environment 237/238. 419-427.
2. Bayramov Z.T. (2002). Devices for Rad.Monitoring and Marine Gamma Survey-ANAA.
3. Agayev F.G. et.al. (2002). Marine gamma survey". ASPG/EAGE International Conference on Petroleum.

OUTLOOK TO NONPROLIFERATION ACTIVITIES IN THE WORLD AND COOPERATION IN PEACEFUL USES OF NUCLEAR ENERGY AMONG TURKEY, CAUCASIAN AND CENTRAL ASIAN COUNTRIES

NEVZAT BİRSEN
Director, TÜDNAEM – Turkish Atomic Energy Agency, Ankara, TURKEY

Corresponding author: n.birsen@taek.gov.tr

ABSTRACT:

On the first call for expanding peaceful uses of nuclear energy, Turkey was one of the first countries to start activities in the nuclear field. Turkish Atomic Energy Authority (TAEK) was established in 1956 and became the member of the International Atomic Energy Agency (IAEA) established in 1957 in same year. TAEK supports, coordinates and conducts the activities in peaceful uses of nuclear energy and acts as a regulatory body and had established cooperation links with developed nations and international organizations. It has established bilateral and multilateral scientific and technical cooperation and signed protocols with scientific organizations of neighboring countries.

Key word: Nuclear energy, research and development.

INTRODUCTION:

Nuclear technology is being widely used in protecting the environment, manufacturing industry, medicine, agriculture, food industry and electricity production. In the world, 438 Nuclear Power Plants are in operation, and 31 are under construction. Nuclear share of total electricity generation have reached to 17 percent. However, 2053 nuclear tests from 1945 to 1999 and 2 atom bombs to Hiroshima and Nagasaki in 1945 have initiated nonproliferation activities aiming to halt the spread of nuclear weapons and to create a climate where cooperation in the peaceful uses of nuclear energy can be fostered.

In addition to international efforts for non proliferation of nuclear weapons, great affords were made for disarmament and banning the nuclear tests which damage the environment. Following the 1st Geneva Conference in 1955 for expanding peaceful uses of nuclear energy, Turkey was one of the first countries to start to conduct research and development in the nuclear field.

Turkish Atomic Energy Authority (TAEK) was established in 1956 and Turkey became a member of the International Atomic Energy Agency (IAEA) established in 1957 by the United Nations for spreading the use of nuclear energy to contribute peace, health and prosperity throughout the world, in the same year.

Turkey is a candidate state to join the European Union and has already signed Custom Union Agreement, also part of the Eurasia Region. So, there are significant developments in the cultural,

M.K. Zaidi and I. Mustafaev (eds.), Radiation Safety Problems in the Caspian Region, 161-164.
© 2004 *Kluwer Academic Publishers. Printed in the Netherlands.*

social, technical, economical and trade relations owing to our common historical and cultural values with the countries in the region and Central Asia.

TAEK was established to support, co-ordinate and perform the activities in peaceful uses of nuclear energy and act as a regulatory body and establish cooperation with other countries and international organizations. In the late 1990's, TAEK, besides building cooperation with various countries, has involved in cooperating with nuclear institutes of Azerbaijan, Kazakhstan, Kyrgyzstan and Uzbekistan for establishment of bilateral and multilateral scientific and technical cooperation in peaceful use of nuclear energy and signed protocols with Academy of Science of Azerbaijan, Nuclear Physics Institute of Kazakhstan, National Academy of Science of Kyrgyzstan and Institute of Nuclear Physics of Uzbekistan Academy of Science.

METHODS:

These protocols enable parties to organize joint projects, conferences, seminars, training programs, establish laboratories for the joint studies and make joint efforts to seek support from their governments and international organizations for these activities. Also, an executive committee has been set up with delegates from each organization under TAEK that also provides the secretarial service for organizing the joint activities. The joint activities carried out are given as follows:

1st Eurasia Conference on Nuclear Science and Its Application - organized in Turkey in the year 2000 by TAEK with coorganizers from the related organizations of Azerbaijan, Kazakhstan, Kyrgyzstan, Uzbekistan and sponsored by IAEA and (OECD/NEA),

2nd Eurasia Conference organized at Almatı in the year 2002 by Nuclear Physics Institute of Kazakhstan with the related organizations of Turkey, Azerbaijan, Kyrgyzstan and Uzbekistan as coorganizers and sponsored by ISTC, SOROS and IDB.

NATO Workshop on Environmental Protection Against Radioactive Pollution organized by TAEK and Nuclear Physics Institute of Kazakhstan at Almatı in the year 2002.

Training Course on Industrial Application of Irradiation Technology organized by TAEK and Academy of Science of Azerbaijan and sponsored by IAEA at Baku in the year 2003.

Joint "Eurasia Nuclear Bulletin" covering activities in peaceful uses of nuclear energy in these countries published yearly since 2002.

CONCLUSION:

Turkey supports the non-proliferation activities that do not prevent the peaceful uses of nuclear energy and in this respect as signed Non-Proliferation Treaty (NPT) and Comprehensive Nuclear Test Ban Treaty (CTBT). Azerbaijan, Kazakhstan, Kyrgyzstan and Uzbekistan have also signed these Treaties following their independence and have become members to IAEA.

STATUS OF SOME COUNTRIES IN EURASIA REGION AND MAJOR COUNTRIES

COUNTRIES	IAEA	NPT	CTBT	COUNTRIES	IAEA	NPT	CTBT
Turkey	1957	1980	2000	Israel	1957	---	1996*
Azerbaijan	2000	1992	1999	Pakistan	1957	-	--
Uzbekistan	1994	1992	1997	India	1957	---	---
Kazakhstan	1994	1994	2002	Armenia	1993	1993	1996*
Kyrgyzstan	2002	1994	1996*	USA	1957	1970	1996*
Turkmenistan	---	1994	1998	UK	1957	1968	1998
Georgia	1996	1994	2002	China	1984	1992	1996*
Tajikistan	2001	1994	1998	France	1957	1992	1998
Iran	1958	1970	1996*	Russia	1957	1970	2000
Iraq	1959	1969	---	Germany	1957	1975	1998

* countries signed, but not ratified.

Acknowledgement: I am thankful to NATO in providing support to present this report at the NATO ARW, Radiation Safety Problems in the Caspian Region, held in 2003, in Baku, Azerbaijan.

REFERENCES:

1. http://pws.ctbto.org/
2. http://www.un.org/Depts/dda/WMD/treaty/
3. www.iaea.org/worldatom/Programmes/Safeguards/sg_protocols.html
4. "2nd International Conference, Problems Related with Non-Proliferation" reports, 14-23 September 1998, Kurchatov-Kazakhstan
5. Nuclear Fuel Cycle and Reactor Strategies: Adjusting to New Realities, International Symposium held in Vienna, Austria, 3-6 June 1997
6. MTR Fuel Cycle Newsletter, No. 14, August 1993

7. Nükleer Silahların Yayılmasının Önlenmesi Anlaşması (Non Proliferation Treaty), N. Birsen, TAEK Rapor No. 1, 1979
8. IAEA INFCIRC / 209 – 254 – 540 –153 – 274 – 225

EMERGING NUCLEAR SECURITY ISSUES FOR TRANSIT COUNTRIES.

I. A. GABULOV
Institute of Radiation Problems of Azerbaijan National Academy of Sciences,
H.Javid Avenue, 31A, Baku, AZ1143

Corresponding author: ibrahim_gabulov@yahoo.com

ABSTRACT:

Nuclear Security and Nonproliferation Issues cover multiple aspects of the security and threat from nuclear terrorism; detection and protection of illicit trafficking of nuclear materials. In the face of emerging threats the prevention of proliferation by the development of effective national system of nuclear export controls is very important for developing-transit countries like Azerbaijan with underdeveloped export controls and strategic locations along trade and smuggling routes between nuclear suppliers and countries attempting to develop nuclear weapons.

Keywords: Nuclear security, developing transit countries, Azerbaijan, nonproliferation, nuclear terrorism, radioactive materials, nuclear materials, radioactive sources, dirty bomb, nuclear export control, illicit trafficking.

INTRODUCTION:

September 11th changed the ways in which the world must view security. The dangers of nuclear terrorism or runaway proliferation are increasingly being developed. Tragic events of September 11 have made these dangers more evident. In the light of increased terrorism preventing the spread of nuclear and nuclear related items as well as radioactive materials that can be used for production so-called dirty bomb is an urgent global claim. Nuclear Security and Nonproliferation Issues cover multiple aspects of the security and threat from nuclear terrorism; detection and protection of illicit trafficking of nuclear materials and other radioactive sources; legal shipment of such type materials as well as nuclear related dual use of these items.

In the face of emerging threats the prevention of proliferation by the development of effective national system of nuclear export controls is very important for Azerbaijan with underdeveloped export controls and strategic locations along trade and smuggling routes between nuclear suppliers and countries attempting to develop nuclear weapons.

The Republic of Azerbaijan has a surface of 86,600 square-km and land borders with Armenia, Georgia, Iran, Russian Federation, and Turkey. To the east the country borders with the Caspian Sea and the coast is about 800 km long. Azerbaijan has a sea border with Turkmenistan and

M.K. Zaidi and I. Mustafaev (eds.), Radiation Safety Problems in the Caspian Region, 165-168.
© 2004 *Kluwer Academic Publishers. Printed in the Netherlands.*

Kazakhstan. Azerbaijan has a population of 8.3 millions and the capital is Baku. The official language is Azerbaijani.

The Republic of Azerbaijan has no nuclear facilities or nuclear materials. Its nuclear activities are limited to typical uses in oil industry, medicine, agriculture and scientific researches. However, Azerbaijan has land and sea borders with countries having nuclear technology, nuclear weapons, nuclear reactors and nuclear materials. It means that nuclear related technology; equipment and materials can be transported both illegally and legally through Azerbaijan's borders.

PROCEDURE:

Nuclear Security and Nonproliferation Issues are very important for the States having nuclear facilities and uranium mining activities. However, the increasing globalization of world trade and open markets in Europe and the United States of America has involved in Nuclear Security and Nonproliferation Issues also transit-countries having underdeveloped controls. Therefore, radioactive or nuclear materials and also nuclear related dual use items can be shipped across international borders and used by terrorist groups. Illicit trafficking of such types of items is a leading proliferation risk today, and efforts to prevent such trafficking have significant weaknesses. Thus, preventing the spread of radioactive, nuclear and nuclear related items is extremely important for the world community and in particular for developing-transit countries. One such country is Azerbaijan. Azerbaijan occupies a very strategic place on the world trade ways as a key element of the Great Silk Way that connects the East and West. Azerbaijan is the "bridge" between Asia and Europe.

Azerbaijan also has extended land borders with the Iran and Russia, and sea border with Kazakhstan, giving the Nuclear Security and Nonproliferation Issues special significance, especially in light of increased terrorist concerns. In this connection it is necessary also to take into account the fact that Russia is at the final stage of the construction of the Busher's Nuclear Power Plant (NPP) and there are plans to construct a few similar NPP on Iran's territory. Moreover, there are certain problems associated with uranium enrichment plant in Natanz. In addition, the issues related to the storage and reprocessing of the spent fuel, which could be used for the manufacture of nuclear explosive devices are not solved.

Thus, in the face of increasing international threat from nuclear terrorism the place and role of Azerbaijan as a transit country has significantly increased. During last years a number of trainings were conducted by the specialists from the U.S. Department of Energy, IAEA and other related organizations for the representatives of both State Border Guard Service and State Customs Committee with the purpose to increase their knowledge and technical level concerning multiple aspects of nuclear security. However, it is not sufficient because Azerbaijan with underdeveloped capabilities for detection and prevention of illicit trafficking radioactive and nuclear materials as well as nuclear related dual use materials and equipment could be used as a transit corridor by terrorists or malevolent groups.

DISCUSSIONS:

Taking into account of all above stated we are able to say that there are the following main problems related to the nuclear security and radiation safety in Azerbaijan:

1. State system for accounting and control of radioactive materials:

- Updating accounting and control for radioactive sources and materials;

- Searching and disposal of "orphan sources" that could be lost by Former Soviet Union military units

2. Control of legally and illegally transport of nuclear related materials and dual-use items through Azerbaijan's borders:

- Monitoring at border control check points by Border Guard and State Customs Committee officers,

- Determination of radioactive or nuclear materials,

- Identification of dual-use items.

In our opinion, first of all, it is need to implement the following steps in order to resolve and improve the existing situation:

- It is necessary to establish effective system for updating existing knowledge of Border Guard and State Customs Committee officers in line with international requirements,

- It is necessary on the basis of international requirements to prepare special training program for Border Guard and State Customs Committee officers,

- It is necessary to conduct regular training courses for Border Guard and State Customs Committee officers,

- It is necessary to establish special program for enhancing and strengthening professional skills of Border Guard and State Customs Committee officers,

- It is necessary to establish the National Center for carrying out detail analysis and expertise of nuclear, radioactive and dual-use items,

- It is necessary to establish special training and research center on the fight against nuclear terrorism.

CONCLUSION:

Azerbaijan Government carries out certain actions in order to upgrade existing situation, but it is very important to assist and share with them modern approach and cumulative knowledge, skills, experience and expertise on multiple issues of nuclear security. Therefore, exchange of information, conducting training courses and other related actions should be done in order to provide assurance that Azerbaijan's borders will be sufficiently guarded from any illegal actions.

REFERENCES:

1. www.iaea.org/world-atom/programs/safeguards/sq-g-protocals
2. SCIENTIFIC TECHNICAL BULLETIN (2000) Scientific - Technical Bulletin. (2000). Scientific Research Technical Center of Federal Border Service of Russia. Moscow.

SOLUTION OF QUESTIONS OF NON-PROLIFERATION IN GEORGIA

ANATOL GORGOSHIDZE
Office of Development of Georgian Border Guards (GBG);
Border Monitoring System of GBG.
State Department of Border Defense
Tbilisi, 380060, GEORGIA.

Corresponding author: anatol_gorgoshidze@hotmail.com

ABSTRACT:

Export Control through Border Monitoring (ECBM) is a challenge for each country, but especially for Newly Independent States (NIS). With the spread of technologies of mass destruction, the monitoring of borders between countries has become the second line of defense against the spread of Weapons of Mass Destruction (WMD), material and technologies for them. Border controls therefore must take into account international attitudes as well as domestic affairs, to preserve the existence of all nations.

Keywords: Export control, border monitoring and weapons.

INTRODUCTION:

Government organizations within NIS needs to have the expertise to develop an ECBM program as they are responsible simultaneously for defending the country and acting as the second line of defense against proliferation. They have tried to find the optimal solution through their own experiments and research [1].

A solution developed by GBG is described here. Georgia is a territorially small transit country and the original function of the transit corridor was to provide a shorter connection for the transportation of goods between Europe and Asia across the Caucasus. During the last ten years, Georgia has built a system of export controls based on its status as a transit state, international treaties and agreements with neighboring countries, domestic legislation and international organizations [2].

The points important to Border Monitoring System:

- Definition of tasks
- Communication subsystem
- Control subsystem
- Observation subsystem

M.K. Zaidi and I. Mustafaev (eds.), Radiation Safety Problems in the Caspian Region, 169-172.
© 2004 *Kluwer Academic Publishers. Printed in the Netherlands.*

In the past ten years, step-by-step, Georgia has formed legislation to address export control. In Georgia, as in many other countries, export control is implemented by the Customs Department of Ministry of Finance and the State Department of the Georgian State Border Defense.

METHODS:

As a transit state, Georgia faced a two-pronged problem in developing its ECBM:

1) Strengthening export controls
2) Developing systems for checkpoints (CP) specifically designed for transit states.

The control of exports in Georgia is not only an international issue, but also a practical and domestic concern. One way for Georgia to develop its economy is to win the trust of companies and the states transporting cargo through the Transport Corridor EUROPE, CAUCASUS, ASIA (TRACECA). The factor of trust is determined, first of all, by safety in the corridor and on Georgia's borders.

Without modern technologies to help supervise and control the check points along the corridor and at the borders, it will be impossible not only to create a Standardized information space but also to ensure the safety and security of cargo according to current requirements. Ensuring the safety and security of cargo in a transit corridor also prevents the proliferation of dual-use technologies.

Effective controls over the transit corridor are not possible without the active development of international legal, economic and political relations; domestic legislation; a high-tech infrastructure; and streamlining of customs and border procedures.

The countries of Georgia, Ukraine, Uzbekistan, Azerbaijan, and Moldova (GUUAM) are cooperating on information pooling for customs and boundary control in the transit corridor. In order to implement information pooling, work must proceed in two directions: Creation of a legal base and a technical base.

As a result of these efforts, we will be able to identify, qualitatively and quantitatively control of cargo, and control of dual-use technologies with greater speed.

Transit legislation in Georgia is based on bilateral and multilateral agreements in which Georgia is one of many participating parties. The most important of these agreements is the multilateral agreement on the development of transportation in TRACECA, Baku in 1998.

The intensive activity in the transit corridor attracts the interest of many criminal and terrorist organizations. While streamlining of border and custom procedures are essential, it should not lead to a weakening of the export control system.

Article 6 in the agreement between custom services of the participants in GUUAM address the necessity of creating and supporting a communication system with the purpose of accelerating data exchange and its protection. Articles 12 and 13 mention issues of confidentiality and use of information. In Kiev, Ukraine, a meeting of GUUAM countries' heads of custom service administrators was held in 2003. This protocol was negotiated and accepted. It mentioned the need to streamline and harmonize custom-boundary procedure in the framework of GUUAM and to create a standardized custom and border information system for GUUAM transit corridor.

Thus, in order to streamline border control procedures, it is necessary to develop a program to implement the information pooling. Priority should be given to weapons of mass destruction nonproliferation and use technology controls. In this way the transit corridor will be able to win the trust of the companies and the states transporting cargo through our country. Thus when comparing the cost of technological equipment for the CPs to loss to the state from criminal activities at the CPs, and political and moral losses, it is easy to see the cost-effectiveness of buying costly technology. In addition to introducing monitoring technology, appropriate data on the cargo must be provided to the CPs before the cargo reaches each point along the transit corridor.

The full spectrum of the political climate, economical situation, and strategic goals of the NIS provides a picture of the tasks involved in building the border monitoring system:

- Create a strategy for dealing with separatist regimes, conflict zones or wars in progress near the border
- Choose the components of the system according to the geographic relief of the country
- Formulate requirements for equipment and military forces for independent field operations and mobile communications connection
- Pursue cost-effectiveness in developing and using the system

The Border Monitoring System is composed of three subsystems: Communication, Control, and Surveillance.

The communications Subsystem and its integration with the other two subsystems lies at the foundation of the monitoring system. Depending on economic, energy and political conditions, the communication subsystem can be built using satellite, relay link, VHF, HF, wire or wireless equipment. The choice of system determines the cost of implementation. For several reasons, the communication subsystem in Georgia is based on HF receiver-transmitter system. One reason was that each unit of the GBG must have military forces with their own independent and mobile field connections. Standardized streamlined reporting forms were developed. Different levels of user interfaces were developed in the Georgian language.

Control Subsystem must have Radiation detector, Computer registration of car number, color, the time of the border crossing on the CP and its own independent power supply for communication unit of Border Post (BP).

In light of the work of CP operators, I cannot overemphasize the importance of regular training for operators, providing them with manual and visual instructions in their native language. It is necessary to provide them with monitoring tools (databases in their native language). This is especially important for dual use products. The operator must be able to compare the name of the company-provider, the transporter, and the end user against the names of companies previously involved in criminal activities. If necessary, the operator also must be able to request data via electronic communications from central servers of the GBG. Also it is necessary to mention that the ability to transmit data in digital format allows timely communication with centralized database so the operator can receive expert preliminary evaluation. The purpose of the Surveillance subsystem is to interrupt the transport of cargo across the border (the borders between CPs).

RESULTS:

The border monitoring system requires a comprehensive approach to the management of the communication subsystem. The system must not only monitor border activities but also coordinate activities of several types of military forces on land, sea, and air.

The Communication Subsystem must include all necessary components for real-time decision-making. All data received from international check points (CP) and border posts (BP) in the form of daily reports or messages about an incident are reflected in the Command Center and be immediately distributed among authorized organizations.

The following activities are performed in order to ensure control over the whole border: continuous collection of information from border CPs, timely criminal activity prevention measures not only outside, but also inside ECBM in order to combat corruption.

CONCLUSION:

Work in ECBM, as for improvement of legislation, personnel training and technology must be an ongoing effort on a permanent basis to make the activities of international organizations of export control more effective. Implementing Export Controls through border monitoring systems is one step in maintaining world order.

REFERENCES:

1. A.Gorgoshidze. (1968). About Several Contemporary Tasks of Technical Safeguarding of Border Agencies of States Members of CIS (Thesis report)".
2. Scientific - Technical Bulletin. (2000). Scientific Research Technical Center of Federal Border Service of Russia. Moscow.

NUCLEAR TERRORISM AND PROTECTION OF ECOLOGY

Zaur Ahmad-zada and Dzhavanshir Aliev.
Institute of Human Rights, National Academy of Sciences
The Baku State University
Baku, AZERBAIJAN.

Corresponding author: bianco_nero@mail.ru

ABSTRACT:

The nuclear terrorism is a difficult phenomenon and has been investigated insufficiently. The special approach is necessary for the decision of this problem. It is necessary to develop the system of effective practical actions on prevention of nuclear terrorism and its consequences for ecology, and also there is a necessity of development of actions on perfection of a legal protection of ecology. All countries of the world have to join actively in struggle against terrorism and consider a question of a legal protection of ecology with big responsibility.

Keywords: Terrorism; ecology; nuclear weapons; protection and environment.

INTRODUCTION:

Recently the terrorism has changed. Before terrorist used to abducted well-known politicians or hijack airplanes, but at the present stage they went on to mass destruction of property and killing of innocent people. The problem of struggle against terrorism in general and against nuclear terrorism in particular is actual problem today. The nuclear terrorism is an opportunity of accomplishment of act of terrorism with use of the nuclear weapons, or accomplishment of explosions on atomic power stations and other objects of atomic energy. Threats to nuclear objects become more and more often and appreciable. It is natural, that accomplishment of acts of terrorism on these objects can lead to ecological catastrophe and can put an irreparable loss to an environment and damage the process of social development. International legal protection of ecology was precisely allocated now in system of the general international law as independent, specific sphere of regulation.

DISCUSSION:

Creation of nuclear weapons in a number of countries, presence of nuclear research reactors, thousands of industrial complex using nuclear material for testing etc, hundreds of power reactors, thousands of nuclear ammunition depots and sets of various objects of an infrastructure, atomic power stations creates objective preconditions for expansion of a field of activity of criminal and terrorist groups and for distributions of possible acts of terrorism.

The nuclear terrorism is meant as an opportunity of accomplishment with use of the nuclear weapons, or accomplishment of explosions on atomic power stations and other objects. Also

M.K. Zaidi and I. Mustafaev (eds.), Radiation Safety Problems in the Caspian Region, 173-176.
© 2004 *Kluwer Academic Publishers. Printed in the Netherlands.*

some experts define nuclear terrorism as application or threat of application of the nuclear weapons or radioactive materials by separate persons, groups or organizations to achieve certain political or economic target. The greatest danger to political and economic system of any state or group of the states is represented with plunder of a nuclear ammunition and threat of its application. Threats to nuclear objects become more and more often and appreciable. Commitment of subversive and terrorist actions can result in heavy consequences concerning objects of the nuclear infrastructure, capable to lead to radioactive contamination of the big territories, sources of water supply, life support systems, etc.

In the world there were some similar cases of threat to nuclear objects from the part of terrorists, and also some cases of use by terrorists of nuclear, chemical and bacteriological substances. For the last few decades in Europe, USA, Russia and Japan, about 200 incidents have taken place which promoted increase of a level of threat. This number included explosions in area of a location of nuclear objects, attempts of penetration on them, kidnapping and murder of nuclear scientists, theft and smuggling of various nuclear materials, etc. As examples, the threat of explosion in Ignalinsk nuclear power plant (Litva, 1994) and three cases that took place in 1995 – France, when terrorists disabled the third energy block of the nuclear power-station in Bleys, Japan - terrorist "Aum Sinrike" applied toxic substances in the underground station in Tokyo, Russia - Chechen extremists hid container with radioactive isotope in Izmailovskiy park in Moscow. Very dangerous situation of the international scale took place in 1961 in France when the group of the generals dissatisfied with politics of the president De Gaulle, intended to grasp nuclear war-head, taking place on test nuclear range in Sahara, and to give Paris- a political ultimatum. Comprehension of a reality of this threat has induced the authorities of France to undertake decisive steps on acceleration of carrying out of test explosion of that nuclear war-head, and the test explosion has been executed in 1961. Wide popularity was received with the act of terrorism that has been accomplished in 1975 in Boston. Group of malefactors have presented the ultimatum to the US authorities about transfer of a large sum of money. Otherwise malefactors threatened to detonate a nuclear warhead. Real danger of explosion was estimated so serious, that the president of the country has been informed about this incident. These cases are evidence of real threat of use by terrorists of the weapon of mass destruction. It is natural, that accomplishment of acts of terrorism on these objects can lead to ecological catastrophe and can put an irreparable loss to an environment and process of social development.

The source of danger can become the nuclear weapons, which have been lost. Now there is very little probability of acquisition of the nuclear weapons by terrorists, but however who can give guarantees that in the near future the situation in this area will not change. Among such losses the powerful American thermonuclear aerial bomb, which was casually dropped in the sea at coast of Spain in 1958, is mentioned. Two nuclear bombs lay at the bottom of Atlantic Ocean near Cape May (New Jersey, USA), which have been dumped from the airplane due to malfunctions arisen during flight. Two more aerial bombs are in water area of Pacific Ocean in area of a Gulf Puget Sound (Washington, USA) and in area of the city of Jurika (Northern Carolina, USA). Also there were some loss of the nuclear weapons in the USSR. The Soviet lost one nuclear submarine in 1986 in Caribbean Sea. There were rockets with nuclear warheads in that submarine. The American authorities have made the big attempt for lifting of this submarine and for extraction of nuclear war-heads, however these attempts have failed. In 1989 in Norwegian Sea, the nuclear submarine "Komsomolets" has sunk. There were torpedoes with nuclear warheads on board of this submarine too. The attempts to lift this submarine or to take torpedoes out also were not successful.

Theft and illegal purchase of split nuclear materials are serious dangers. The structure of the nuclear weapons is not secret, and the elementary nuclear charges can be created even in primitive conditions. Scientists warn, that it is the most real at use of enriched uranium as weapons nuclear substance for creation of a nuclear charge. Such charge can be created with a high level of reliability without carrying out of nuclear tests and that provides reserve and suddenness of its military use. The charge of such design "Little boy" has been blown up in Hiroshima without carrying out of preliminary tests. The opportunity of similar development of events, in particular, is confirmed with one case that took place in 1975. The student of one of the American universities has created the working scheme of the explosive nuclear device of the elementary type on the basis of the data published in an open press. According to the expert opinion, if there was a nuclear substance in that construction in case of its operation there would be a nuclear explosion. In such a case, such explosion may be carried out by the terrorists, it will lead to the big ecocatastrophe.

Threat of nuclear terrorism to humanity and its consequences for ecology last years is aggravated because terrorism as the phenomenon of political and social struggle from an internal problem of the separate states turns to the serious international factor. First of all the danger of nuclear terrorism is inextricably related with expansion of scale of activity of the various extremist and fundamentalist religious groups considering terror as powerful means for achievement of purpose. Comprehension of depth of the danger that connected to nuclear terrorism and its threat of ecology, has special requirements to maintenance of highly effective system of nuclear safety.

Up to the middle of 60th years of 20th century, protection of an environment was not put forward as an independent political problem, and its scientific substantiation has not been developed enough as a diversified, complex, global problem. Only dynamical development in 70-80th years of scientific bases of global problems has allowed to allocate the rules of law concerning to protection of an environment, into special group.

International legal protection of ecology was precisely allocated now in system of the general international law as independent, specific sphere of regulation. The principle of inadmissibility of radioactive pollution of environment covers both military, and peaceful use of nuclear power. Formation and the statement of this special principle of international law of environment take place in two ways - contractual and usual, with observance by the states of existing international practice.

Azerbaijan Republic takes a special place at the turn of Europe and Asia. In territory of Azerbaijan geopolitical interests of many countries are crossed. Except for that the Caucasus, one of which important components is Azerbaijan, represents very difficult and specific region.

Serious threat for the region is the Armenian Atomic Power Station (APS). As it has been marked above, accomplishment of acts of terrorism on nuclear objects can lead to ecological catastrophe and can put an irreparable loss to an environment and process of social development. In a case of accomplishment of act of terrorism on the Armenian APS, there can be a big ecological catastrophe in the region from which the inhabitants of Armenia, Azerbaijan, Turkey, Iran, Georgia and the countries of the Near East suffer.

Azerbaijan is relatively young independent state which is integrated into the world community. Azerbaijan is the full member of many international organizations. Also Azerbaijan takes a special geopolitical place in region and Azerbaijan is on crossing of Europe and Asia. Consequently, the Azerbaijan cannot be away from world process and has to conduct active struggle against terrorism, and also has to conduct a correct politics in the field of a legal protection of ecology. Considering all aforesaid, Azerbaijan should improve the legislation in the field of the ecological rights and especially in the field of the struggle against terrorism in view of new displays of terrorism, namely in view of probability of accomplishment of acts of terrorism with use of the weapon of mass destruction, according to experience of the advanced countries. Perfection of the legislation in these areas should conduct on the basis of observance of specificity of region and answer the international conventions.

As to struggle against terrorism, for today the parliament of Azerbaijan ratified the international convention of the United Nations "About struggle against bombing terrorism" and also the international convention "About struggle with financing of terrorism" which has been accepted by General Assembly of the United Nations (1999) [1,2]. In the internal legislation of the Azerbaijan Republic there is a Law "About struggle against terrorism" [3].

As to a legal protection of ecology and preservation of the environment, for today the parliament of Azerbaijan ratified about 20 international normative-legal acts (the conventions, reports and so on) in the field of ecology and preservation of the environment. In the internal legislation of the Azerbaijan Republic there is a number of laws on protection of ecology and an environment, in particular: "The Law on Ecological Safety" (1999); "The Law on Preservation of the Environment" (1999) [4]; "The Law on Especially Protected Objects and Territories of the Nature" (2000) [5]. The work in the field of the ecological rights and struggle against terrorism, and also perfection of legislations of the Azerbaijan Republic in these areas is one of priority tasks of lawmaking of the country at the present stage.

CONCLUSION:

The nuclear terrorism is a difficult sociopolitical, economic, religious and military-technical phenomenon and need to have action from all countries of the world to join hands to struggle against terrorism.

REFERENCES:

1. "Acting International Law", M., 1997;
2. "Development of international law of environment", A. Timoshenko, M., 1986;
3. "Nuclear terrorism in the modern world", V. Belous, "CISR", M, 1999;
4. "Ecological law", M. Brinchuk, M., 1999;
5. "Bulletin of the parliament of Azerbaijan Republic", 2003.

THERMOLUMINESCENCE PERSONAL AND MEDICAL DOSIMETRY

MÁRIA RANOGAJEC-KOMOR
Ruđer Bošković Institute, P.O.Box 180, 10002, Zagreb Croatia

Correspoding author: marika@irb.hr

ABSTRACT:

Thermoluminescence dosimetry (TLD) is applied worldwide for personal and medical dosimetry. In this paper a brief summary of theoretical background and measurement procedure of TLD is presented. The main characteristics of various TLD systems (batch homogeneity, sensitivity, reproducibility, linearity, light sensitivity, fading and energy dependence) are described mostly from the viewpoint of use in personal and medical dosimetry. Some remarks about organisation and trends in personal dosimetry and some special medical dosimetry applications (in paediatric X-ray diagnostics and computerised thomography) of TLDs are also reviewed.

Keywords: Thermoluminescence dosimetry, personal dosimetry and medical dosimetry

INTRODUCTION:

Dosimetry can be defined as the measurement of dose of ionizing radiation using an adequate instrument. Quantitative dose measurements of ionizing radiation enable the establishment of a numerical relation between the dose and physical, chemical and biological changes induced by this dose.

Personal dosimetry is the determination of exposure from external radiation sources of personnel working in the field of ionizing radiation. Medical or clinical dosimetry in this work means the determination of dose to patients during a medical treatment. Personal and medical dosimetry are usual regulated by the corresponding national laws. The most commonly used passive integral personal dosimeters are film and thermoluminescence (TL) dosimeters. In routine personal dosimetry ionisation chambers are also used sometimes. They enable the direct reading of dose or dose rate.

The basis for radiation protection is the exact determination of doses delivered to personnel or patients. The knowledge of irradiation dose enables, if necessary, additional radiation protection measures, or in case of accidents to decide the method of medical intervention. Usualy, personal dosimeter is placed on the left side of the chest under the radiation protection clothing. In special working processes dose of radiation can be measured on different parts of the body, such as the extremities, eyes, gonads etc. In medical dosimetry radiation doses have to be measured on various parts of the body on patients during various diagnostic or therapeutic medical treatment.

Dosimeter systems for personal and medical dosimetry have to fulfil the following requirements (1):

M.K. Zaidi and I. Mustafaev (eds.), Radiation Safety Problems in the Caspian Region, 177-190.
© 2004 *Kluwer Academic Publishers. Printed in the Netherlands.*

- dosimeter has to be small, simple to carry, mechanically stable and cheap,
- it has to be able to measure the radiation dose in the range from several $\propto$Sv to several Sv,
- the energy dependence of the dosimeter system has to be low or it has to be taking into correction,
- the dose response has to be independent on the position of the dosimeter,
- gamma and X radiation has to be measured separately from the beta and neutron radiation,
- fading of the dose must be small or known to enable correction,
- environmental conditions (humidity, light, temperature etc.) should not influence the uncertainty of dose measurements,
- the evaluation of the dosimeter has to be rapid, simple and cheap.

According to these requirements film and TL dosimeters are the most suitable for personal dosimetry. For patient dose measurements TLDs are more convenient because of their small sizes. In this paper an overview of TL dosimetry for these applications is given. The mechanism and characteristics of the TL dosimetry systems have been studied intensively over many years. Despite the substantial results reported over these years, there are additional questions and the novel systems remaining for investigations.

Thermoluminescence dosimetry (TLD) is nowadays worldwide used for personal and environmental monitoring (2) as well as for dose measurements on patients in the medical applications. The large number of the papers related to personal and clinical dosimetry at the last three Solid State Dosimetry Conferences (3-5) indicate the importance of this method. Valuable books and monographs deal with the theory and application of thermoluminescence (6-17).

PRINCIPLES OF TL DOSIMETRY

Following the absorption of the energy from ionizing radiation, molecules of the irradiated substance come from ground state (g) to one of the excited states (e). These excited molecules can loose their energy on various ways; one possibility is light emission, i.e. luminescence. The prompt light emission is called fluorescence (Fig. 1a), while delayed emission is phosphorescence (Fig. 1b).

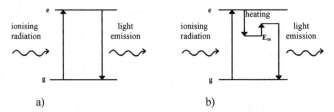

Figure 1. Schematic illustration of fluorescence (a) and phosphorescence (b)

Phosphorescence is the return of an electron from excited state to the ground state delayed by the metastable energy level (E_m). Thermoluminescence (TL) is kind of phosphorescence when heating of the material stimulates the return of an electron trapped in the metastable level. Many crystalline materials show TL characteristics after irradiation with ionizing radiation. TL detectors are natural or synthetic materials emitting light during heating of the sample, previously excited. The intensity of light is proportional to the dose of irradiation. The intensity of emitted light as a function of heating temperature is called the thermoluminescence glow curve (Fig.2).

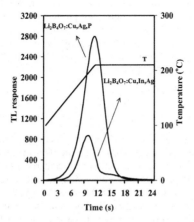

Figure 2. Typical glow curve of $Li_2B_4O_7$ detectors irradiated with $D\gamma$= 2 mGy (^{137}Cs source). Reader: TOLEDO 634. Heating rate: 10 K/s

The heating rate is constant, and therefore the glow curve is often shown in the literature as the TL intensity as a function of time of the measurement cycle. In this case the time-temperature profile (T) in addition to the glow curve can be very useful. A typical glow curve of two $Li_2B_4O_7$ detectors (18,19,20) - which was developed first of all for personal dosimetry applications - can be seen in Fig. 2. The integrated area under the glow curve between two temperature values or the peak height are used for dose calculation. For dosimetry purposes they have to be linearly proportional to the radiation dose.

TLD SYSTEM

A TL dosimetry system contains the following components:
- TL detectors
- TL reader
- measurement cycle.

All these components influence the characteristics and uncertainty of the system and therefore their optimization is always very important.

TL detectors

Numerous materials such as LiF, CaF_2, Al_2O_3, MgB_4O_7, $CaSO_4$ and $Li_2B_4O_7$ show thermoluminescence characteristics. To satisfy the requirements of personal, medical and other applications they contain different activators, which influence the dosimetric characteristics of the detector. The most commonly used detectors are the LiF detectors produced by Harshaw (recently Saint Gobain Crystals and Detectors, http://www.bicron.com/): LiF:Mg,Ti is known as TLD-100. Its isotope variations with Li^6 and Li^7 (TLD-600 and TLD-700) are also used, first of all in mixed field (n+γ) dosimetry. CaF_2 doped with Mn and Dy as well as the high sensitive Al_2O_3:C and LiF:Mg,Cu,P (21) are also very often used. LiF:Mg,Cu,P appears under various commercial names depending on manufacturer and production procedure (TLD-700H (22), GR-200, G-200A (23,24), MCP-N (25)). The advantage of LiF detectors for personal dosimetry is their tissue equivalence and therefore their small energy dependence. On the other hand the new highly sensitive materials are suitable for measurements of very low doses (several ∝Sv). Recently efforts are done to develop $Li_2B_4O_7$ based personal TL dosimeters (18,19,20) and Si doped LiF dosimeters (26,27).

TL readers

The TL readers are instruments which enable the reproducible heating of TL detectors in various physical forms (powder, chips, discs etc.) from ambient temperature to 300-400 °C, the collection of the emitted light by photomultiplicator and recording of data. Today there are several commercial TL readers (Harshaw, Panasonic, Vinten, Teledyne etc.) available. Many laboratories developed their own reader (28,29,30,31,32). For routine personal monitoring automatic readers are convenient, because they enable the evaluation of large number of dosimeters in a relative short time. For research purposes manual readers are better choice, because they are more flexible in variety of the reading parameters.

Measurement cycle

The measurement cycle contains the following steps: annealing, package and storage, irradiation, readout and mathematical evaluation. All these steps influence the end dose value; therefore they all have to be well defined for an optimum measurement cycle.

Annealing is a pre-irradiation heat treatment to achieve reproducible results. It enables the re-use of TL detectors because retains the characteristics after repeated irradiations. The way of annealing (heating rate, temperature, cooling rate) influences significantly the sensitivity of TL detectors, therefore identically performed annealing is needed for standardization of the detector sensitivity. The package serves for protection against environmental influences (light, humidity, mechanical damage), for insurance of electronic equilibrium during irradiation as well as for compensation of energy dependence by using adequate filters. After packaging the detector becomes a dosimeter. The following step of the measurement cycle is the irradiation of dosimeters because of two purposes: one serves for calibration usual by ^{226}Ra, ^{137}Cs, ^{60}Co sources, the other is the irradiation in the field, which have to be measured. TL dosimetry is not absolute method, therefore the dosimeter has to be calibrated in a known irradiation field to determine the calibration factor. During the field irradiation in personal and clinical dosimetry the dosimeter is located on various parts of the body of the personnel or the patients. The readout of the irradiated dosimeters is the heating of the detector with a constant heating rate to a preciously defined maximum temperature. The result of the readout is the glow curve. The readout value or TL response is the area under the glow curve or the peak height. The mathematical evaluation enables the calculation of the evaluated value from the readout value using calibration factor, correction factors (reader, fading, energy, sensitivity, background) and the algorithm to express the dose in terms of interest (33). In personal dosimetry nowadays the evaluated value is mostly expressed in terms of $H_p(10)$ (34).

CHARACTERISTICS OF TLD SYSTEMS

Thermoluminescence can be observed in many materials, however only a few of them fulfil the requirements of TL dosimetry. In Table 1 the requirements for personal and medical dosimetry are shown (14,35). According the recommendations of the International Atomic Energy Agency (IAEA) the uncertainty in personal dosimetry for very low doses can be even 100% compared to the dose equivalent of an ideal dosimeter worn at the same point of the body at the same time (35).

The main characteristics of a TLD system are: batch uniformity, linearity, sensitivity, reproducibility, fading and energy dependence.

The batch uniformity shows the differences in sensitivities of individual detectors within the batch. It is expressed as one standard deviation of the mean TL response after irradiation each detector in the batch with the same dose.

Table 1. Requirements for TLD systems in personal and clinical dosimetry

Task	Dose range (Sv)	Uncertainty SD (%)	Tissue equivalency
Personal dosimetry extremity, whole body, tissue	$10^{-5} - 5 \times 10^{-1}$	-30, +50	important
Clinical dosimetry Radiotherapy X-ray diagnostics	$10^{-1} - 10^{2}$ $10^{-6} - 10^{1}$	±3.5 ±3.5	very important important

SD: standard deviation

For linearity in personal dosimetry the IEC Standard (33) requires that the response should not vary by more than 10% over the range of 0.5 mSv to 1 Sv. In ideal situations the dose response function is linear in a broad dose range. At high doses, in practically all cases, supralinearity is possible. At very high dose, in the range of saturation, the dose response function may even decrease in some cases. In Fig. 3 the dose response functions for some detectors are shown (36) expressed as the measured doses in comparison with the nominal doses of irradiation. The R^2

values of linear fitting for all detectors in Fig. 3 are 0.999 that means very good linearity for the detectors investigated.

The sensitivity of a TL material is defined as the ratio of the TL response and absorbed dose. This parameter has not an absolute value for a certain TL material because it depends on many factors such as the kind and concentration of activators, system of readout, heating profile etc. Therefore the relative sensitivity is used i.e. the TL response of a certain dosimeter is compared very often with the TL response of TLD-100.

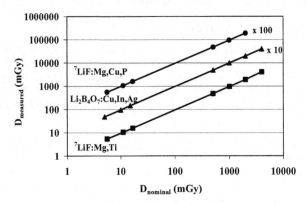

Figure 3. The linearity of the dose response function of some TL detectors

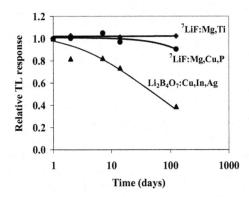

Figure 4. Fading of various detectors (36) (source: ^{137}Cs, D = 2 mGy)

Dosimeters with adequate sensitivity has to be choosen according to the purpose of dosimetry. For personal dosimetry mostly LiF:Mg,Ti detectors (TLD-100, MTS-N) are used. In addition to LiF:Mg,Ti for patient exposure measurements in X-ray diagnostics high sensitivity detectors such as LiF:Mg,Cu,P or LiF:Mg,Cu,Na,Si are recommended while for radiotherapy less sensitive detectors, such as $Li_2B_4O_7$:Cu,In are convenient.

The reproducibility of the complete TLD system is very important. It can be defined in various ways (2,33) and is influenced by many components of the TLD system (detector, reader, irradiator etc.). The reproducibility values of LiF:mg,Ti (TLD-100), LiF:Mg,Cu,P (GR-200A) and LiB_4O_7:Cu,In,Ag expressed as 1 standard deviation of calibration factors (36,37) are ±4 %, ±5 % and ±11 %, respectively. The detectors were evaluated by the same reader (TOLEDO 654), irradiated on the same ^{137}Cs source (1 mGy-15 mGy). In general ±10 % reproducibility is acceptable for personal dosimetry while for clinical dosimetry, especially in radiotherapy better reproducibility is required.

The fading means the decrease of TL signal in function of time after irradiation. Fading characteristics of TL materials have to be always investigated under different conditions (temperature, humidity, light effects). The fading can be improved in different ways (2), such as thermal cleaning procedure, 24 hours storage between irradiation and readout, mathematical correction etc. Fig. 4 shows the fading of various detectors in 125 days after

The energy dependence is the change of TL response at a certain dose as a function of the energy of the radiation and is very important in personal and clinical dosimetry. Therefore it is essential that the absorption and scattering of radiation in the dosimeter would be similar to that in the material in which the dose have to be measured, i.e. in the tissue. That means that the detector material has to be tissue equivalent, i.e. the effective atomic number (Z_{eff}) of the detector should be close to 7.4 (Z_{eff} of tissue). It was shown in the literature (37,38,39) that the energy absorption characteristics of TLDs depend not only on the effective atomic number of the respective TLD but also on the influence of any added dopant. The calculated and measured energy dependences of some detectors are shown in Fig. 5 (17). The differences in energy dependence of LiF detectors with various activators compared to the conversion coefficient from air kerma (K_a) freein-air to $H_p(10)$ are shown in Fig.6 (36,39).

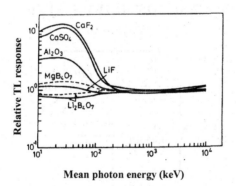

Figure 5. Calculated (—) and measured (- -) energy dependence of various TL detectors

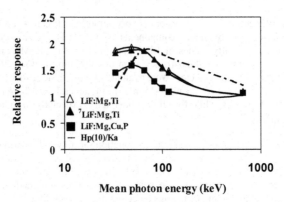

Figure 6. Energy dependence of various LiF TL detectors compared to $H_p(10)/K_a$ (Sv/Gy). Relative response: dose measured on phantom relative to the delivered dose K_a.

The personal dosimetry should be well characterized and approved by the international or national standard performance test programs. With appropriate choice of the dosimetry system, (especially the detector and the heat treatment) very wide dose range can be measured (several μGy -10^5 Gy).

5. PERSONAL DOSIMETRY – ORGANIZATION, INFRASTRUCTURE, TRENDS:

The Croatian Radiation Protection Institute was establish in 1997 to maintain records on radiation sources and persons handling these sources, to participate in formulating the radiation protection laws and regulations and to organize training for occupational staff.

PERSONAL DOSIMETRY SYSTEM

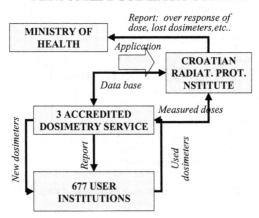

Figure 7. The Croatia System of personal dosimetry services.

From 677 user institutions 523 are health, 107 industrial, 26 veterinarian, 11 research and 10 others institutions. The 3 accredited dosimetry services controlled 4793 individuals working in these 677 institutions in 2002. About 80% of these staff works in medical institutions. About 1200 persons are controlled by TLD, the others by film dosimeter. The dose distribution of all controlled persons in year 2002 (Table 2) shows that nobody exceeded the annual dose limit of 50 mSv (41).

Table 2. Dose distribution of the controlled persons in Croatia in year 2002.

Dose (mSv)	Number of individuals
<0.1	4004
0.1 – 1	638
1 – 15	151
Total:	4793

The advantages from a central database of personal dosimetry are the followings:

- the supervisory authority will be informed promptly about any over response of individual dose,
- the collected results enable the analysis of protocol of the dosimetry services and their competition help them to improve the dosimetry service,

- in case of changes or winding-up of any dosimetry service the data base keep its data,

- the data are controlled and actual.

The most significant problem in this way of organization rise from the attitude of some users to the dosimetric control which cause a delayed return of relative large number of dosimeters (20-30%). To improve this symptom adequate and permanent education and stronger control of the authorities is needed.

The European Radiation Dosimetry Group, EURADOS (http://www.eurados-online.de/), established a working group, which aims at assisting in the process of harmonization of individual monitoring as part of the protection of occupationally exposed persons, in the framework of EC Directive (40). The working group carried out two projects. In the first project (42) the working group prepared a catalogue of dosimetric systems able to assess $H_p(10)$ and a consolidated performance test to investigate the dosimetric quality of systems for assessing $H_p(10)$. In the second, recently ongoing project (43) the working group addresses the following issues:

- An overview of the national and international standards and other documents of relevance that are of importance for the quality

- An inventory of methods and services for assessing the dose due to external radiation and of direct and indirect methods for assessing the dose due to internal contamination.

- An inventory of new developments in individual monitoring (electronic dosimeters)

- An inventory of problems of non-dosimetric origin in individual monitoring that impair the quality of the dose assessment

Among the services participated 31 use film dosimeters for photons and beta rays and 61 use TLDs. An end product of the second project will be the generation of the "Eurados Database of European Dosimetric Data" containing all the information collected from facilities of 28 European countries collaborating with EURORADOS. A periodic update of the Database will be carried out in the future to improve the amount of information collected, trying always to reflect the actual situation of the dosimetry in Europe.

CLINICAL DOSIMETRY – SOME EXAMPLES

It is well known that among man-made radiation sources, the highest contribution to the population exposure is due to the radiation used for medical purposes (Fig. 8).

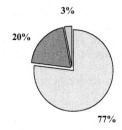

77% **Natural radiation**
20% **Medical irradiation**
3%   **Other sources of radiation**

Figure 8. Mean yearly effective equivalent doses from natural and man-made sources of radiation

Medical irradiation comes from X-ray diagnostic treatments, such as conventional radiology, computerized thomography (CT), from radiotherapy irradiations and procedures in nuclear medicine. The thermoluminescence dosimetric systems have been found to be adequate for measuring doses on the patients after the characterisation of TLD systems described above. The exact knowledge of doses delivered to different organs or to the whole body enables:

- to estimate the exposure to the population
- to estimate the risk
- to specify radiation protection measures
- to assess the justification of the procedure in terms of overall patient benefit and safety measures.

Since 1969 the IAEA and the World Health Organization (WHO) has performed regularly postal TLD audits to verify calibration of radiotherapy beams in different radiotherapy units (44). The main purpose of the program is to provide an independent verification of the dose delivered by radiotherapy treatment. The program has checked about 4000 clinical beams all over the world. In many instances significant errors have been detected in the beam calibration. Subsequent follow-up actions have helped the radiotherapy centers to resolve the discrepancies and to correct the dosimetry procedures, thus preventing further mistreatment of patients.

The determination of the doses received by patients in X-ray diagnostics is complicated because very low doses at low and variable energies has to be measured and there exists considerable variation in radiation doses delivered to patients (different X-ray equipment, different staff, etc.). For example a well-tested TL system irradiated with X-rays in Secondary Standard Dosimetry Laboratory show ± 4 % reproducibility in 3 identical measurement cycle (Table 3). For comparison, the standard deviation of the same TL system in the irradiation field of an X-ray unit in a hospital without patients (±21 %) and on children patients (±100 %) during chest examination indicate the real situation (38).

Table 3. Reproducibility of LiF:Mg,Ti based TLD system

X-ray irradiation	Number	Energy of radiation	SD (%)
in SSDL*	3 irradiation	80 keV	± 4
in the field of routine X-ray unit (without patients)	10 irradiation	90 kV	± 21
on patients	10 patients	variable	± 100

*SSDL: Secondary Standard Dosimetry Laboratory (3 irradiation)

The X-ray examinations of thoracal organs occupy the first place in medical exposure, particularly in childhood.. Therefore radiation doses have been determined on various parts of the body on 50 children of five different age groups during thorax and sinus examinations. At standard projections of lung examinations (posteroanterioral (PA) and lateral chest radiography), the highest radiation doses have been measured (38,45) on the back (0.26 ± 0.11 mSv). The doses on sternum and on gonads were 0.20 ± 0.09 mSv and 0.04 ± 0.02 mSv, respectively. When examining the paranasal cavities (46) highest radiation doses have been found on the left eye (0.14 ± 0.05 mSv). The large standard deviation (up to 100 %), even within one age group, has been caused by a series of factors (the irradiation conditions, different X-ray equipment, patients' physical characteristics, work methods etc.). Some of these factors cannot be influenced upon in everyday work. To evaluate the real exposure of children to radiation, it is necessary to carry through a statistically relevant number of measurements.

Using the measured mean dose values (45) and the risk factors (47) the risks of leukemia, lung cancer and genetic damages in X-ray diagnostics of respiratory tract of children were estimated.

Though the evaluated risks was not alarming, all patient protection measures should be carried out. The dosimetric measurements proved also the efficiency of various radiation protection methods.

Table 4. Risk estimation of radiation damage in X-ray diagnostics of respiratory tract of children

Damage Organ	Leukaemia Red marrow	Cancer Lungs	Genetic Gonads
Risk factor* (10^{-2}Sv^{-1})	1.04	0.80	1.33
Weight factor w_T	0.12	0.12	0.20
Skin dose (mSv)	0.26	0.20	0.04
Risk**	3.081.854	5.208.333	9.398.496

*Number of children examined cause damage for 1 patient

From X-ray diagnostic treatments the highest doses to the patients result from CT examinations. CT accounts for 2-3% of all X-ray examinations and it contributes about 20% of the population doses. The distribution of surface radiation doses during CT of the head is shown in Table 5 (48). The surface radiation received by the thyroid, chest and gonads results from scattered radiation. The ratio of the minimum to the maximum of the measured surface dose varies by a factor of 4 to 7. This is understandable when we consider that the dose varies according to the position, size and shape of the patients' body; that the distance of the primary X-ray beam from the organ. The number of scans significantly varies from one patient to another. Similar dose variations were found in New Zealand, Sweden, and the United Kingdom and in Japan (49) measured even on a hypothetical average adult phantom.

Table 5. Distribution of surface radiation doses during CT of the head

Surface dose on	Lowest value (mGy)	Highest value (mGy)
Right eye lens	6.44	70.92
Left eye lens	6.45	71.10
Thyroid	0.63	2.86
Chest	0.25	1.04
Right gonad	0.09	0.57
Left gonad	0.05	0.38

CONCLUSIONS:

It is necessary to undertake all kinds of measures to minimize the harmful consequences of radiation during exposure to man made sources. For this purpose most national laws/regulations require permanent personal dosimetry control of staff working in the ionizing radiation field.

Considering, radiation protection of patient, an optimum factor for X-ray examinations should be decided by image quality but also the absorbed dose. This aim can be achieved by frequent dosimetry control and permanent education of personnel. In radiotherapy the verification of dose delivered to patients is the best radiation protection measure.

Frequent measurements of radiation dose in medicine facilitate the control of equipment; identify equipment requiring additional safety measures; facilitate patient risk assessment; aid in directing efforts toward reducing total radiation doses. The TLD system is suitable for personal and medical dosimetry as the detectors are small, simple to carry, mechanically stable and cheap, the evaluation of the dosimeters is rapid and simple.

After the choice of a TLD system with adequate characteristics for personal and clinical dosimetry (sensitivity, reproducibility, linearity and energy dependence) has been made, careful performance test and calibration has to be carried out according to national and international standards in order to fulfill the uncertainty and accuracy requirements for personal and medical dosimetry (33,35,40). To achieve and maintain international standards, intercomparisons of TLD systems are highly recommended.

Acknowledgement: The author is very grateful to Saveta Miljanić for her contribution to the paper. NATO provided full financial support to present this paper at the NATO Advanced Research Workshop, Radiation Safety Problems in the Caspian Region, BAKU.

REFERENCES:

1. Ranogajec-Komor, M.,Vekić, B. (2002). Personal Dosimetry in: Radijacijske ozlijede, Dijagnostika i liječenje. - in Croatian. Zagreb. Medicinska naklada. 53-59.
2. Ranogajec-Komor, M. (2003). Thermoluminescence Dosimetry-Application in Environmental Monitoring. Radiation Safety Management. 2, 2-16
3. Pető, Á., Uchrin, Gy. (Editors) (1996). Solid State Dosimetry . Radiat. Prot. Dosim. 66.
4. Delgado, A., Gómez Ros, J. M.(Editors). (1999). S.S.D. Radiat. Prot. Dosim. 85.
5. Horowitz, Y. S., Oster, L.(Editors). (2002). Solid State Dos. Radiat.Prot.Dosim. 101.
6. Daniels, F. et.al (1953). Thermoluminescence as a Research Tool. Science 117, 343-349.
7. Cameron, J.R.et.al (1968). Thermoluminescent Dosimetry. MU Wisconsin Press.
8. Becker, K. (1973). Solid State Dosimetry. Boca Raton, Florida. CRC Press.
9. McKinlay, A.F.(1981). Thermoluminescence Dosimetry. M.P.Handbooks5, B.A. Hilger.
10. Oberhofer, M. et.al. (1981). Applied Thermol. Dosimetry. ISPRA Courses. B.A. Hilger.
11. Horowitz, Y. S. (1984). Thermoluminescent Dosimetry. Boca Raton, Florida. CRC Press.
12. McKeever, S. W. S. (1985). Thermoluminescence of Solids. Cambridge University Press.
13. Vij, D. R. (1993). Thermoluminescent Materials. New Jersey. PTR Prentice-Hall.
14. McKeever, S. W. S., Moscovitch, M., Townsend, P. D. (1995). Thermoluminescence Dosimetry Materials: Properties and Uses. Ashford, Kent. Nuclear Technology Publishing.
15. Chen, R.et.al. (1997). Theory of Thermol. and Related Phenomena. World Scientific.
16 Furetta, C. et.al. (1995). Operational Thermoluminescence. World Scientific Publishing.
17 Uchrin, Gy., Ranogajec-Komor, M. (1988). Thermoluminescence Dosimetry, in Ionizing Radiation, Protection and Dosimetry. Boca Raton, Florida. CRC Press. 123-132.
18. Prokić, M. (2001). Lithium Borate Solid TL Detectors. Radiat. Meas. 33, 393-396.
19. Prokić, M. (2002). Dosimetric Characteristics. Radiat. Prot. Dosim. 100(1-4), 265-268.
20. Furetta, C. et.al. (2001). Dosimetric Characteristics.N.Inst.Meth.Phys.Res.A 456, 411-417.
21. Boss, A.J.J. (2001). High Sens. Thermol.Dosimetry. N.Inst Meth.Phys.Res. B 184, 3-28.
22. Luo, L. Z.et.al.(2002): Evaluating Extremity Dosemeters. Radiat. Prot. Dosim. 101, 211-216.

23. Zha, Z.et.al(1993). Preparation of LiF:Mg,Cu,P Radiat. Prot. Dosim. 47, 111-118.
24. Tang, K.et.al.(1999). An Improved LiF:Mg,Cu,P Chip. Radiat. Prot. Dosim. 84, 227-229.
25. Bilski, P. (2002). Lithium Fluoride: LiF:Mg,Ti to LiF:Mg,Cu,P. RprotDos 100,199-206.
26. Budzanowski, M. et.al. (2001). Dosimetric Properties.Radiat. Measur. 33, 537-540.
27. Kaiyong Tang. The Study of a New LiF:Mg,Cu,Si TL Chips. Rad.P.Dos. to be published
28. Dražič, G. et.al.(1986). The Influence of TL G.Curve.Radiat. Prot. Dosim. 17, 343-346.
29. Szabó.et.al. (1977). The TLD-04B Thermoluminescent Reader.KFKI Report, KFKI-77-33.
30. Van Dijk, et.al.(1990). Performance Analysis of the TNO TLD. R.Prot. Dosim. 34, 171-174.
31. Wang, J. Q.et.al.(1993). A TLD Reader with Chip Micropocessor. N.Tr.Rad.M.22, 893-895.
32. Apáthy, S., Deme, S., Fehér, I. (1996). TLD System. Rad. Prot. Dos. 66, 441-444.
33. International Electrotechnical Commission. (1991). Thermol. Dosimetry Systems for Personal and Environmental Monitoring. CEI/IEC International Standard 61066:1991-12
34. International Commission on Radiation Units and Measurements. (1998). Determination of Dose Equivalents Resulting from External Radiation Sources. Bethesda, MD. Report 39
35. IAEA (1999). Asssessment of Occupational Exposure Due to External Sources of Radiation.IAEA Safety Standards Series No RS-G.1.3, IAEA Safety Guide, 17-24.
36. Knežević, Ž. (2004). Characterisation of Themoluminescence Dosimeters in the Fields of Photon Radiation of Different Energies. Master's Thesis, Zagreb. Ruđer Bošković Institute.
37. Miljanić, S., Ranogajec-Komor, M., Knežević, Ž., Vekić, B. (2002). Main Dosimetric Characteristics of Some Tissue-Equivalent TL Detectors. Radiat. Prot. Dosim. 100, 437-442.
38. Ranogajec-Komor, M., Muhiy-Ed-Din, F., Milković, Đ., Vekić, B. (1993). Thermoluminescence Characteristics of Various Detectors.Radiat. Prot. Dosim. 47, 529-534.
39. Miljanić, S., Knežević, Ž., Štuhec, M., Ranogajec-Komor, M., Krpan, K., Vekić, B. (2003). Energy Dependence of New TLDs in Terms of $H_P(10)$ Values. Rad.Prot. Dosim. 106, 253-256.
40. The Council of European Union. (1996). Laying Down Basic Safety Standards for the Protection of the Health of Workers, Official Journals of European Communities, 159.
41. Kubelka, D., Sviličić, N., Kralik Markovinović, I. (2003). Dosimetric Control of Personal in Zone of Ionising Radiation. Proc. of the Fifth CRPA. pp. 177-182– in Croatian.
42. Lopez, M.A.et.al (2003). Harmonisation . Radiat. Prot. Dosim. 105, 653-656.
43. Lopez, M.A.et. (2004). Harmonisation of Individual Monitoring in Europe. Proc. of 11th International Congress of the IRPA.Madrid, Spain, 23-28. May 2004. –to be published
44. Izewska, J. et.al (2002). TLD Postal Dose Audit Service.Rad.Prot. Dos.101, 387-392.
45. Milković, D., Ranogajec-Komor, M., Krstić-Burić, M., Hebrang, A. (1991). Mit Thermolumineszenz- Rontgenaufnahmen von Kindern und Jugendlichen. At.Lung.17, B67-B72.
46. Milković, D., Ranogajec-Komor, M., Krstić-Burić, M., Hebrang, A. (1993). Bestimmung der Bestrahlungsdosis in radiologischer Diagnostik paranasaler Sinusse bei Kindern und Jugendlichen. Atemw.-Lungenkrkh., 19, S101-S104.
47. ICRP (1991). Recommendations of the ICRP - ICRP Publications 60, NY Pergamon Press.
48. Vekić, B., Ranogajec-Komor, M. (1996). Determination of Patient Surface Doses from Computerized Tomography Examinations of the Head. Rad.Prot. Dosim. 66(1-4), 311-314.
49. United Nation Scientific Committee on the Effects of Atomic Radiation. (1993). Sources and Effects of Ionizing Radiation. UNSCEAR 1993 Report to General Assembly. NY. UN.

AQUEOUS-PHASE CHEMICAL REACTIONS IN THE ATMOSPHERE

ALEXANDER N. YERMAKOV, IGOR K. LARIN, ANTON A. UGAROV
Institute of Energy Problems of Chemical Physics
Russian Academy of Sciences
Leninsky pr. 38, Bldg. 2, 119334, Moscow, RUSSIA.

Corresponding author: ayermakov@chph.ras.ru

ABSTRACT:

An importance of aqueous-phase chemical reactions is demonstrated on an example of sulfur dioxide oxidation in the atmosphere. This process results in acid rains with known negative impact on the environment. Till now such an aqueous-phase process in the atmosphere remained not fully understood. Their dynamics and mechanism in non-precipitating troposphere clouds is studied with a coupled gas-phase and aqueous-phase chemical model.

Keywords: Environment, ecology, human health and contamination.

INTRODUCTION:

Energy production strongly impacts on the environment in particular, to the increase of the atmospheric loads of acidic gases and particle loading e.g. to the acid rain environmental problem [1,2]. Scientific evidence has shown these depositions can harm ecosystems and human health [3] At present, the main stream of atmospheric chemistry research efforts is focused around a new field, namely the chemistry of atmospheric heterogeneous and multiphase reactions with the purpose of studying of its potential impact on the environmental [4,5]. During the past two decades great strides have also been made toward understanding aqueous-phase chemical reactions in cloud droplets, in particular, those involved transition metal ions (TMI) [6]. The knowledge how these metal ions accelerate chemical reactions in the droplet phase will help to develop and evaluate a relevant role of TMI in the atmosphere and probably others aqueous-phase atmospheric processes [7]. In contrast to a fairly extensive knowledge of gas phase reactions in the troposphere, current understanding of cloud chemistry yet not well determined. Despite their low water volume content in the gas (typically 10^{-7} vol/vol), the droplets provide an ideal reaction media to occurrence of chemical reactions in the atmosphere especially, which involve TMI, which might act as efficient catalysts, and hence too much more efficient aqueous phase reaction pathways then the respective uncatalyzed gas phase reactions. Many aspects of this problem are still far from certain.

Uncatalyzed depletion of SO_2

An analysis within a coupled gas-phase and aqueous-phase chemical model has shown that the droplets accelerate significantly sulfur dioxide oxidation in the atmosphere. Its rate removal from the gas increases by a factor of ◻ 10^2 compare with that provided by only gas-phase chemical

M.K. Zaidi and I. Mustafaev (eds.), Radiation Safety Problems in the Caspian Region, 191-195.
© 2004 *Kluwer Academic Publishers. Printed in the Netherlands.*

reactions, see curves "gas" and "gas-droplet" in Figure. Modeling of the gas-liquid processes revealed that the SO_2 is depleted through "fast" and "slow" modes. First mode is caused by liquid phase reaction between hydrogen peroxide that present initially in the gas and bisulfite anions arising in the droplets due to dissolution of sulfur dioxide and its ionization $SO_2 + H_2O$ □ $HSO_3^- + H^+$. The characteristic time of fast mode is of □ 10^3 s. High solubility of hydrogen peroxide favors the transferring of gaseous H_2O_2 into the liquid phase [3]. The ionization of sulfur dioxide in the droplets is another factor favoring the mode of SO_2 removal from the gas. The autocatalytic reaction [8] between H_2O_2 and HSO_3^- ($H_2O_2 + HSO_3 + H^+$ □ $2H^+ + SO_4^{2-} + H_2O$) keeps also an importance during the "slow" mode of SO_2 depletion, see Figure. Contrary to that occurred in the "fast" mode, this reaction in a course of the "slow" mode involves the hydrogen peroxide that is produced in the droplet phase. Its production is provided due to the uptake of HO_2 radicals from the gas followed recombination $HO_2 + HO_2/O_2^-$. In addition, some portion of SO_2 is depleted by means of parallel autocatalytic reaction [9] with in-droplets produced HSO_5^-: ($HSO_5^- + HSO_3^- + H^+$ □ $3H^+ + 2SO_4^{2-}$). These oxidizing species in turn arise in the droplets due to the uptake of OH radicals from the gas: $OH + HSO_3^-$ □ $\overset{\text{$O_2$}}{\text{□}}$ SO_5^- □ $\overset{HO_2}{\text{□}}$ HSO_5^- [10].

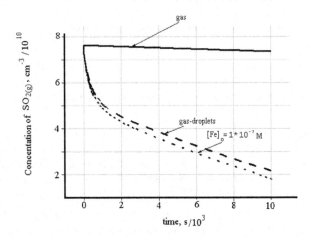

Fig. Box model of the atmosphere with a coupled gas-phase and aqueous-phase chemical reactions. Liquid water content 10^{-6}, droplet radii 1 $\propto$ P_{SO_2} = 3 ppbV, $P_{H_2O_2}$ = 1 ppbV, temperature 298 K. The calculated time histories of SO_2 depletion for different scenarios. Upper curve "gas" represents gas-phase "dry" process. Curve marked by "gas-droplets" corresponds to the oxidation in the presence of droplet phase. Bottom curve gives the evolution in time of SO_2 when gas-liquid process occurs in the presence of iron with metal ion content of 10^{-7} M.

Iron-catalyzed oxidation of SO_2

These ions influence SO_2 removal from the gas in the "slow" mode, see Fig. In daytime the impact of these ions on sulfite oxidation is initiated by uptake of a sum of OH/HO_2 radicals from the surrounded gas and also due to absorption of UV light [11]. To specify these effects, the basic knowledge of iron-catalyzed sulfite oxidation ("dark" process) is needed [12]. The catalytic process, as was recently found [13], represent a chain reaction with degenerated branching. Perhaps most importantly, the incorporation of the HSO_5^- + Fe (II) step into the regeneration of catalytically active ferric ions which does not deplete its role over the iron redox cycle. The radical-radical recombination $SO_5^- + SO_5^-$, which terminates the cycling between ferric and ferrous ions, represents a gross but not a net loss of the chain-carriers, because nearly all of them are reformed through the branching step $HSO_5^- + Fe(II) \rightarrow Fe^{2+} + H_2O + SO_4^-$, $SO_4^- + HSO_3^-$ → →² → SO_5^- in just a few seconds or somewhat longer. A branching mechanism is thus the only possible means of allowing the catalytic process to reach a stationary state. Observations that may be considered as evidence (fingerprints) of rate variations in sulfite depletion due to the branching mechanism were explored in details and the related dynamics of the chain-carriers and metal ions cycles were discussed within frameworks [14]. In particular, the most important is found to be the aspect related to the intrinsic limitation of the cycle of metal ions. This limitation governs the extent of the oxidative/reducing potential of sulfite solutions with respect to the Fe (III/II) couple, thereby governing the quasi-state partioning between ferric and ferrous ions. Such a view enables examination of those conditions under which the limitation to the rate of the catalytic reaction is controlled by the reduction or reoxidation of ferric ions. Readily applicable kinetic criteria and kinetic diagrams to delimit the conditions were given. In such a framework, the majority of known anomalies of the catalytic reaction receive an explanation. The scheme shown below demonstrates both chain-carriers' and metal ions' cycles during iron-catalyzed sulfite oxidation under laboratory conditions named below as a "dark" process.

"DARK" PROCESS

Chain channel of sulfite depletion	Catalytic channel of sulfite depletion
$Fe(III) + HSO_3^- \;\square\; Fe(II) + H^+ + SO_3^-$	$2Fe(III) + 2HSO_3^- \;\square\; 2Fe(II) + 2H^+ + 2SO_3^-$
$SO_3^- + O_2 \;\square\; SO_5^-$	$2SO_3^- + 2O_2 \;\square\; 2SO_5^-$
$SO_5^- + HSO_3^- \;\square\; HSO_5^- + SO_3^-$	$SO_5^- + Fe(II) + H^+ \;\square\; HSO_5^- + Fe(III)$
$HSO_5^- + Fe(II) \;\square\; Fe(III) + OH^- + SO_4^-$	$HSO_5^- + Fe(II) \;\square\; Fe(III) + OH^- + SO_4^-$
$2HSO_3^- + O_2 \;\square\; OH^- + H^+ + SO_3^- + SO_4^-$	$2HSO_3^- + 2O_2 \;\square\; OH^- + H^+ + SO_4^- + SO_5^-$

Our chemical model [15] indicated that sulfite oxidation in the atmosphere is accelerated compare with that found in the laboratory (in a course of "dark" process). The reason is the acceleration of iron redox cycling due to uptake of HO_2 from the gas. Iron ions present in the droplet phase even in extremely low amounts of 10^{-7} M are capable to catalyze the conversion of the radicals: $HO_2 \;\square\; ^{Fe}\square\; SO_5^-$. In details the mechanism of the conversion is shown:

ATMOSPHERIC PROCESS

$$FeOH^{2+} + HO_2 \;\square\; Fe^{2+} + H_2O + O_2$$
$$Fe^{2+} + HSO_5^- \;\square\; FeOH^{2+} + SO_4^-$$
$$SO_4^- + HSO_3^- \;\square\; SO_3^- + H^+ + SO_4^{2-}$$
$$SO_3^- + O_2 \;\square\; SO_5^-$$
$$\overline{HO_2 + HSO_3^- + HSO_5^- \;\square\; SO_4^{2-} + H^+ + H_2O + SO_5^-}$$

Iron present in the droplet phase can be considered as an agent accelerated production of the intermediated HSO_5^-. It has been found that the rate of the $HO_2 \xrightarrow{Fe} SO_5^-$ conversion under conditions of interest is limited by the rate of the reaction $FeOH^{2+} + HO_2$. As a result the balance between Fe (III) and Fe (II) shifts to ferric ions. The conversion $HO_2 \xrightarrow{Fe} SO_5^-$ accelerates significantly the rate of SO_5^- production in the droplets. These in turn form HSO_5^- thus accelerating the rate of SO_2 removal from the gas; one SO_2 is removed per two HO_2 radicals captured by the droplets, whereas the only one captured OH provides the consuming of one SO_2. The obtained results agree with available recent field observations of the balance between Fe (III) and Fe (II) over the episodes of so-called radiation fog or orogrpaphic clouds.

Manganese-catalyzed oxidation of SO_2

The known row of the catalytic activity of transition metal ions, namely Mn > Fe > Cu > Co does not correlate with the known row of their redox potentials [13]. Our analysis of all available works related to Manganese (II)-catalyzed sulfite oxidation led to a conclusion that the only iron ions have the catalytic activity under laboratory conditions ("dark" process). Manganese ions being itself catalytically inefficient are capable the only enhance the catalytic effect of iron ions [14]. This situation dramatically changes when passing from laboratory to atmospheric conditions. In the atmosphere the fluxes of OH/HO_2 radicals can directly ($HO_2/O_2^- + Mn (II) \rightarrow Mn (III)$) or indirectly ($OH + HSO_3^- \rightarrow SO_5^-$; $SO_5^- + Mn(II) \rightarrow Mn(III)$) activate manganese ions thus accelerating sulfite depletion and the rate of SO_2 depletion from the gas. The obtained results related to the activation of manganese ions by direct and indirect manners are discussed. In addition some results are given to illustrate the effect of iron-manganese co-catalysis of sulfite oxidation in the atmosphere (so-called synergism Fe/Mn [16].

CONCLUSION:

Detailed knowledge of cloud processes is essential to forecast of real distances from a source of the pollutants to the location where they can actually be deposited onto the earth surface.

Acknowledgement: This work is supported by Russian Fund of Basic Researchers (Grant 01-02-16172). Presented at the NATO ARW held at BAKU, Azerbaijan in September 11-14, 2003.

REFERENCES :

1. Regens J.L, Rycroft R.W. The Acid Rain Controversy. (1988). Bayker and Taylor's, London.
2 Gail S., Acid Rain, Overview Series, Our endangered planet, (1992), Lucent Book. Inc.
3. Seinfeld J.H.,.Pandis S.N.(1986). Atmospheric Chemistry and Physics. Wiley, NewYork.1326.
4. Molina M.J., L.T.Volona and C.E.Kolb; Gas-phase and heterogeneous chemical kinetics of troposphere and stratosphere. Ann. Rev. *Phys. Chem.* 47 (1996) 327-367.
5. Learning to Manage Global Environmental Risks. (2001). The MIT Press, Cambridge, USA.

6. Mirabel V., G.A.Salmon, R.van Eldik, C.Winckier, K.J.Wannowius, and P.Warnek. (1996). Review of the activities of the EUROTRAC Subproject HALIPP. Berlin. Springer. 7-74.

7. Graedel T.E, M.L.Mandich,C.J.Weschler. (1986) Kinetic model studies of atmospheric droplet chemistry.J. Geophys. Res. 91 (1986) 5205-5221.

8. Betterton E.A., M.R Hoffman. (1988) Oxidation of Aqueous SO$_2$. J.Phys.Chem. 92. 5962 - 5965.

9. Jacob D.J. (1986). Chemistry of OH in remote clouds and its role in the production of formic acid and peroxomonosulfate, J. Gephys. Res. 91. 9807-9826.

10. Herrmann H., B. Ervens, H.-W. Jacobi, R. Wolke, P. Nowacki, R. Zellner. (2000). CAPRAM: A Chemical Aqueous Phase Radical Mechanism. J. Atmos.Chem. 36. 231–284.

11. Martin L.R, M.W. Hill, A.F.Tai, T.W.Good. (1991). The iron catalyzed oxidation of sulfur (IV) in aqueous solution AND effects of organics at high and low pH, J.Geoph.Res. 96. 3085-3097.

12. Warnek P, J.Ziajka. (1995). Reaction mechanism of the iron (III)-catalyzed autoxidation of bisulfite in aqueous solution: steady state description for benzene. Ber.Buns.Phys.Chem. 99. 59-65.

13. Brandt Ch, R.van Eldik.(1995). Transition metal-catalyzed oxidation.Ch.Rev. 95. 119-190.

14. Yermakov A.N, I.K.Larin, A.A.Ugarov and A.P. Purmal. (2003). Iron-catalyzed oxidation of sulfite: From established results to a new understanding, Prog.Reac.Mech. 28. 189-255.

15. Yermakov A.N, I.K.Larin, A.A.Ugarov, A.P.Purmal. (2003). Atmospheric oxidation of SO$_2$ catalyzed by iron ions, Kinetika and Catalysis, (Rus). 44. 524-537.

16. Ibusuki T, K Takeuchi. (1987) Sulfur dioxide oxidation. Atmos.Environ. 21. 1555-1560.

MEASUREMENT OF NATURAL RADIOACTIVITY IN PHOSPHOGYPSUM BY HIGH RESOLUTION GAMMA RAY SPECTROMETRY.

H. YÜCEL*, H. DEMIREL*, A. PARMAKSIZ*, H. KARADENIZ*, İ. TÜRK ÇAKIR*, B. ÇETINER*, A. ZARARSIZ*, M. KAPLAN*, S. ÖZGÜR*, H. KIŞLAL**, M. B. HALITLIGIL** AND İ. TÜKENMEZ**

* Turkish Atomic Energy Authority (TAEA)-Ankara Nuclear Research and Training Center, 06100 Beşevler, Ankara, Turkey

** TAEA-Ankara Nuclear Research Center in Agriculture and Animal Science, 06983 Saray, Kazan, Ankara, TURKEY.

Corresponding author: haluk.yucel@taek.gov.tr

ABSTRACT:

Phosphatic fertilizers are manufactured from phosphate ores which contain naturally occurring radionuclides such as ^{238}U and its daughter products. Large amounts of phosphogypsum is obtained and has ^{226}Ra. The natural radioactivity was measured by a gamma ray spectrometer. The activity concentration ^{226}Ra was 465 Bq/kg as a mean value. The natural external gamma radiation dose rate is estimated to be 205 nGy/h at 1 m above ground level when 1-1.5 kg of phosphogypsum is applied in 1 m^2 surface area of field.

Keywords: Radioactivity, exposure, food, fertilizer, phosphoric acid, radon.

INTRODUCTION:

Phosphatic fertilizers are manufactured from the industrial processing of rock phosphate ores which are known to contain naturally occurring radionuclides such as ^{238}U and its daughter products. A huge amounts of phosphogypsum (PG) is obtained as a by-product, causes serious storage and environmental problems around phosphoric acid industries. Spreading of PG to the land surface is the most commonly adopted disposal and used in many countries. It may enhance the soil aggregation which promote the invasion of beneficial fauna causing better media to crop growth and improve the soils with fertility problems. One problem with using PG in agriculture is that it contains radioactive radium and associated radon emanation. In the late 1980's, the U.S Environmental Protection Agency (EPA) has permitted the controlled use of PG in agriculture.

Phosphogypsum (PG) is an important by-product of phosphoric acid and fertilizer industry. It consists of gypsum (CaSO$_4$2·H$_2$O) and contains some impurities such as phosphate, fluoride, organic matter and alkalis. The composition of PG varies depending upon the source of rock phosphate and the process for manufacturing phosphoric acid [1].

M.K. Zaidi and I. Mustafaev (eds.), Radiation Safety Problems in the Caspian Region, 197-204.
© 2004 *Kluwer Academic Publishers. Printed in the Netherlands.*

The presence of radionuclides puts restrictions on the use of PG in soil amendments and in building materials. Also, the impurities of phosphogypsum (PG) seriously restrict the industrial use of PG in cement industry as a retarder [2]. However, the PG currently produced in Turkey is generally stockpiled near the production sites, for example, 10 million tones of the wet PG material in Samsun Fertilizer Industry accumulated between 1970 and 1973. Furthermore, huge amounts of PG from Mersin, İskenderun, Bandırma (BAGFAŞ) and Yarimca (İzmit) fertilizer facilities are either distributed onto soil or discharged into sea.

Many problems are associated with the disposal of PG on the soil, such as the movement of fluorate, sulphate, and radionuclides potentially affecting water sources, and the exhalation of radon from PG stacks that could affect the local workforce and residents in the areas near the stacks. In addition to these problems, there is the need of an extensive physical space for PG deposition because the large quantities of PG are continuously being produced.

The characterization of radioactive elements should be the first step in trying to evaluate the radiological impact associated with the use of PG. Hence the primary objective of this work is to characterize the naturally occurring radionuclides in Turkish PG, aimed at providing a database for assessment of environmental radiological impacts due to the subsequent uses of this material. Also it is important to complete the lack of data on the radioactivity contents of PG produced in Turkey.

Therefore, within a research contract between Turkish Atomic Energy Authority and Turkish Soil & Fertilizer Research Institute, the studies related to the measurement of radioactivity contents of PG and its agriculture use for soil amendment are now being conducted. The results of [226]Ra measured in PG samples taken from a stock near Iğdır, located in North- East of Turkey, are presented in this paper.

PROCEDURE:

Sampling

About 60,000 tones of PG stock in Iğdır will probably be used in soil improvement in this region. Sixty PG samples were collected from this stock representing the PG of Samsun Phosphoric Acid and Fertilizer Production Facility. The PG samples were grinded, powdered and dried at 110°C in an oven, and then sealed in the plastic containers for gamma ray spectrometric measurements.

Elemental analysis of PG by X-Ray Fluorescence Spectrometry
Elemental analysis of PG were carried out by using an energy dispersive X-ray fluorescence spectrometer. It uses a Si (Li) detector with a resolution of 190 eV at 5.9 keV. The compact Oxford ED2000 EDXRF system is a tube excited spectrometer in which the power of tube is 50 W and its maximum current is 1000 mA, and X-rays are originated from a Rh target. The results of elemental analysis of PG are given in Table 1.

Gamma Spectrometric Measurements in PG:

Commercially available two p-type HpGe detectors were used in the radioactivity measurements. One has a relative efficiency of 12.4% and the resolution of ~2 keV for 1332.5 keV of ^{60}Co. The other has a relative efficiency of 10% and the resolution of 1.8 keV for 1332.5 keV. The energy and efficiency calibration of the detectors used were carried out according to the standard methods for detector calibration and analysis described in ASTM-E181 standard [3]. The γ-ray counting setup was calibrated with uranium reference standard of known U content in an identical geometry. The photopeaks used for ^{226}Ra in the γ-ray spectra of PG samples were 609 keV of ^{214}Bi and 351 keV of ^{214}Pb. For the ^{228}Ra determination a photo-peak of 911 keV (^{228}Ac, half life of 6.15 h) was used, the results obtained for ^{228}Ra were lower than minimum detectable activity (MDA) of the present γ-counting system. For the ^{232}Th determination, a photopeak of 583 keV (^{208}Tl, short-half life of 3.05 min.) was used and ^{40}K activity is determined from 1460 keV peak.

Table 1. Elemental Analysis of PG by XRF method.

Major elements	Concentration (g/kg)
Ca	219.4
S	197.8
Si	5.3
P	2.3
Al	1.6
Fe	1.2
F	Not determined
Minor elements	Concentration (mg/g)
K	255.5
Ba	129.5
Ti	116.2
Sr	90.0
Y	81.3
Zn	65.3
Cu	28.0
Cr	23.0
Ni	16.8
Zr	10.0
Mo	4.4

RESULTS & DISCUSSION:

Many authors studied the PG material obtained in North America, South America, Europe and Africa [4]. The interval for ^{226}Ra specific activity in PG given in literature are normally 15 to 2000 Bq/kg and are given Table 2.

Table 2. [226]Ra activity of PG materials produced in various countries in the world [4]

Continent	State/ Country	[226]Ra (Bq/kg)
North America	Florida	907
	Louisiana	1100
	Alberta	910
South America	Brazil	273-591
Europe	Netherlands	800
	Spain	508
Africa	Tanzania	3219
	Morocco	1420

The specific activities of [226]Ra in PG samples ranged from 283 ± 8 Bq/kg to 586 ± 12 Bq/kg (as a mean value of 465 ± 15 Bq/kg). [232]Th activity was of the order of 10 Bq/kg and the [40]K activity was 70 Bq/kg. Each of them is given as a mean value. The radioactive equilibrium that normally exists between [238]U and its daughters or [232]Th and its daughters, is broken during the wet chemical process by reaction of phosphatic rocks with sulphuric acid. Hence, the radionuclides are partitioned depending on their solubility into phosphoric acid or PG. The measured radioactivity of Turkish PG taken from the stock in Iğdır are given in Table 3.

Table 3. The specific activities of [226]Ra, [232]Th and [40]K contained in PG in Iğdır.

	Ra-226			Th-232			MDA	K-40			MDA
	Activity (Bq.kg^{-1})			Activity (Bq.kg^{-1})				Activity (Bq.kg^{-1})			
PG-01	283,1	±	7,5	10,3	±	1,2		< MDA			32,9
PG-02	454,4	±	8,7	< MDA			9,8	< MDA			32,4
PG-03	416,7	±	8,8	< MDA			10,2	< MDA			30,1
PG-04	411,4	±	8,8	< MDA			10,6	< MDA			30,0
PG-05	448,9	±	8,0	< MDA			6,6	79,4	±	14,0	
PG-06	399,4	±	8,0	< MDA			9,2	< MDA			27,3
PG-07	434,4	±	9,3	< MDA			10,9	< MDA			33,6
PG-08	397,9	±	8,0	< MDA			7,2	89,9	±	14,5	
PG-09	389,2	±	8,6	< MDA			7,0	78,5	±	20,3	
PG-10	376,8	±	7,9	< MDA			9,3	< MDA			29,2
PG-11	384,1	±	8,7	12,4	±	4,6		59,9	±	20,9	
PG-12	432,5	±	8,9	< MDA			11,1	< MDA			33,0
PG-13	406,1	±	9,0	< MDA			7,2	67,6	±	21,9	
PG-14	460,9	±	9,8	< MDA			6,7	64,9	±	21,6	
PG-15	481,9	±	10,0	< MDA			12,2	< MDA			37,9

PG-16	450,8 ± 9,9	< MDA	7,4	65,5 ± 22,3	
PG-17	482,4 ± 12,4	< MDA	12,0	< MDA	31,5
PG-18	370,2 ± 7,9	< MDA	5,5	116,6 ± 16,1	
PG-19	348,9 ± 7,8	32,8 ± 8,2	< MDA		29,5
PG-20	398,8 ± 12,1	< MDA	12,3	< MDA	33,1
PG-21	389,8 ± 9,3	< MDA	13,0	< MDA	36,2
PG-22	384,8 ± 8,3	< MDA	5,8	82,3 ± 14,5	
PG-23	456,2 ± 11,9	< MDA	11,6	< MDA	30,7
PG-24	426,0 ± 8,7	< MDA	6,2	72,5 ± 18,8	
PG-25	432,6 ± 9,2	< MDA	11,1	< MDA	27,5
PG-26	491,0 ± 10,2	< MDA	7,1	68,4 ± 21,0	
PG-27	503,3 ± 9,7	< MDA	11,9	< MDA	32,3
PG-28	499,4 ± 10,0	< MDA	7,0	75,4 ± 22,8	
PG-29	533,1 ± 10,6	< MDA	7,4	60,4 ± 33,8	
PG-30	559,1 ± 12,6	21,0 ± 6,9	< MDA		31,6
PG-31	587,8 ± 11,9	15,6 ± 4,3	< MDA		37,7
PG-32	487,0 ± 9,9	< MDA	8,3	76,2 ± 25,6	
PG-33	483,9 ± 10,0	< MDA		< MDA	
PG-34	518,6 ± 12,0	< MDA	10,6	< MDA	30,5
PG-35	473,2 ± 8,6	< MDA	7,2	92,3 ± 15,4	
PG-36	473,2 ± 9,5	< MDA	10,7	< MDA	29,6
PG-37	537,8 ± 9,9	< MDA	10,9	< MDA	32,4
PG-38	452,1 ± 8,6	< MDA	7,0	77,0 ± 13,3	
PG-39	438,5 ± 10,8	< MDA	11,3	< MDA	29,8
PG-40	464,2 ± 8,6	< MDA	7,5	72,9 ± 23,6	
PG-41	445,7 ± 10,5	< MDA	10,1	< MDA	29,8
PG-42	496,9 ± 9,1	< MDA	8,2	81,7 ± 17,0	
PG-43	499,4 ± 9,8	< MDA	12,6	< MDA	37,7
PG-44	474,5 ± 8,8	< MDA	5,7	63,0 ± 16,0	
PG-45	478,2 ± 9,3	< MDA	10,9	< MDA	29,8
PG-46	514,0 ± 11,6	< MDA	11,8	< MDA	34,2

PG-47	586,1 ±	11,5	< MDA			13,3	< MDA	36,1
PG-48	459,9 ±	9,1	< MDA			10,7	< MDA	28,6
PG-49	538,4 ±	10,2	20,8	±	7,2		< MDA	34,8
PG-50	485,1 ±	9,9	< MDA			11,7	< MDA	34,3
PG-51	505,8 ±	10,0	< MDA			11,6	< MDA	32,4
PG-52	541,2 ±	9,8	< MDA			13,3	< MDA	35,9
PG-53	515,7 ±	9,2	< MDA			12,2	< MDA	32,3
PG-54	469,5 ±	9,2	< MDA			10,3	< MDA	31,4
PG-55	538,7 ±	10,1	< MDA			14,4	< MDA	37,4
PG-56	451,2 ±	9,1	< MDA			10,4	< MDA	31,0
PG-57	583,2 ±	10,1	< MDA			12,2	< MDA	35,9
PG-58	489,2 ±	9,5	< MDA			10,8	< MDA	33,2
PG-59	520,7 ±	9,2	< MDA			12,0	< MDA	35,8
PG-60	512,4 ±	9,5	< MDA			10,5	< MDA	28,0

MDA: Minimum Detectable Activity.

HULL and BURNETT [5] confirmed that in Florida the majority of ^{228}Th and ^{230}Th is incorporated into phosphoric acid whereas the ^{226}Ra, ^{210}Pb and ^{210}Po into PG. However, in this work, the data on the phosphate rock/PG mass ratio is not available, because the origin of phosphate rock used is not known. It is therefore the partitioning of radionuclides during the P_2O_5 acid production has not been studied.

Fernandes et.al. [6] confirmed that up to 94.1% of thorium in the phosphatic rock was incorporated into phosphoric acid, thus the PG samples presents a low level of this radionuclide as well as the present results for thorium were obtained very low than expected.

As Ra belongs to the alkaline earth metal group, its chemical behavior is similar to Ca, therefore it is incorporated into the PG during the chemical processing of the phosphatic rock. Hence, ^{226}Ra is the major source of radioactivity in PG. When the phosphatic rock contains large quantities of ^{232}Th, the PG will have ^{228}Ra. One of the environmental worries with radium is due to the fact that it follows the same biological pathway as calcium allowing incorporation into living organisms. ^{226}Ra is preferentially accumulated in the bone. In addition, ^{226}Ra is transformed through α-decay in ^{222}Rn, a noble gas with a short half-life of 3.82 d.

Therefore, it is important to evaluate the radiological impact of ^{222}Rn and its daughters within the PG, because radon gas can exhale from the PG stacks containing ^{226}Ra. In the late 1980's, the U.S Environmental Protection Agency (EPA) has permitted the controlled use of PG in agriculture if the ^{226}Ra levels are less than 370 Bq/kg (10 pCi/g). The importance given to ^{226}Ra is illustrated by the maximum limit of 2664 Bq/m^2/h established by EPA for the exhalation rates of ^{222}Rn from decommissioned PG stacks [7].

Dose rate calculations:

UNSCEAR has given the dose conversion factors for converting the activity concentrations of ^{238}U, ^{232}Th and ^{40}K into doses (in n·Gy/h per Bq/kg) as 0.427, 0.662 and 0.043, respectively [8]. Using those factors the total absorbed gamma dose rate calculation in air at 1 m above the ground level is calculated as following:

$$D \, (n·Gy/h) = 0.427 \, C_U + 0.662 \, C_{Th} + 0.043 \, C_K$$

Where, C_U, C_{Th}, C_K are the specific activities of U, Th and K in PG. From the specific activities of ^{226}Ra, ^{232}Th, ^{40}K in PG taken from the stock in Iğdır given Table 3, the total gamma dose rate was calculated as 205 n·Gy/h (about 0.2 μSv/h). In this dose estimation, the conversion factor of 0.427 for ^{238}U was assumed to be equivalent with that of ^{226}Ra.

The world average of total absorbed gamma dose rate due to the presence ^{238}U, ^{232}Th and ^{40}K in soil is 43 n·Gy/h (0.043 μSv/h) as reported by UNSCEAR [8].

The calculated extra gamma dose rate result for PG is about 4-5 times higher than the world average gamma dose rate in soil. However, when a 1-1.5 kg of PG is applied to the soil of 1 m^2 surface area, the exhalation rates of ^{222}Rn due to phosphogypsum are estimated to be between 465 Bq/m^2/h and 698 Bq/m^2/h. The radon gas exhalation rates estimated are lower than the maximum limit of 2664 Bq/m^2/h established by EPA.

CONCLUSION:

Disposal of PG has the potential to elevate natural levels of uranium series nuclides. The measured activities of ^{226}Ra in the PG stock in Iğdır as an averaging value of 465 Bq/kg is higher than the maximum limit of 370 Bq/kg established by EPA by 25%. This result implies that the use of the PG in agriculture needs to be taken some protective measures. However, this material may be used in soil amendments by employing some suitable methods, such as dilution or extraction of radium from PG, provided that the maximum limit of ^{226}Ra activity would be kept below 370 Bq/kg.

Acknowledgement: This work was supported by Turkish Atomic Energy Authority and Turkish Soil and Fertilizer Research Institute under research contract no. AP.2.B.2. and NATO provided financial support to present this paper at the NATO ARW 2003, Baku, Azerbaijan.

REFERENCES:

1. Mays, D.A. and J.J. Mordvedt, (1986). Environmental Quality, 15. 78-81.
2. Kumar, S. (2002). Construction & Building Materials 16. 519-52.
3. Annual Book of ASTM Standards, (1993). Vol. 12, 02, E181 1-28.
4. Silva, N.C. et al. (2001). J. Radiaoanal and Nucl. Chem. 249 (1). 251-255.
5. Hull, C. D., Burnett, J. (1996). Environ Radioactivity, 32 (3). 213.

6. Fernandes, E.A.N., Fukuma, H. T., Quinelato A.L., (2000). Radiochimica Acta, 88 (09/11) 809.
7. Code of Federal Regulations, (1998), Title 40, Vol.7 parts 61.202 and 61.204 (40 CFR 61.202 and 40 CFR 61.204)
8. UNSCEAR (United Nations Scientific Committee on the Effects of Atomic Radiations. (1988). Sources, Effects and Risks of Ionizing Radiation. Report to the General Assembly, with Anexes, United Nations, New York.

ELECTRET POLYMER MATERIALS FOR DOSIMETRY 0F α-IRRADIATION

A. M. MAGERRAMOV, M.K. KERIMOV, E.M. HAMIDOV
Institute of Radiation Problems, National Academy of Sciences
Az 1143, H. Javid av. 31-A, Baku, Azerbaijan

Corresponding author: elsevar@mail15.com

ABSTRACT:

The structural and technological aspects of development of electrets, piezoelectric polymer thin films (PPTF) in dosimetry of α-irradiation at wide range of absorbed doses of radiation are analyzed. The different ways of obtaining high effective thin electroactive polymer films and new composite materials are considered. The possible mechanism of increasing parameters of a dosimeter is discussed.

Keywords: electret, polymer thin film, radiation dose and composite material.

INTRODUCTION:

Polymer films (polyetylene (PE), polytetrafluorethylene (PTFE), polypropylene (PP), polystyrene (PS)) are widely applied in dosimetry of α-radiation. The dose, D, are estimated relative to the changing of determined properties of the polymer under action of radiation [1,2]. The development of radiation technology, methods of environment monitoring, radiation ecology, radiobiology and other aspects of radiation chemistry and physics require new tasks in dosimetry [3-6]. It is known, that the polymer film dosimeters and ionizing radiation indicators are more sensitive at high doses of radiation at range of $D = 1 - 10^8$ Gy [5,6]. Extension of these ranges are measured with different dosimeters and transducers, having various principles of work based on thermoluminescence, semiconductive, membranous materials and gases. The main requirements presenting to them are high sensitivity to different types of radiation, extension of lower and upper limits of measured doses, precision and reproducibility of results. One of the important problems of gamma and electron radiation dosimetry is the determination of small doses of radiation, which are necessary for processes of radiation sterilization technology and for estimation of biological and ecological actions to organisms and plants [3,5,7].

METHODS:

Solutions to the problems indicated above can be achieved with both new radiation sensitive materials and systems developed with early applied materials which underwent modification. It is necessary to note, that the access to development of dosimetric devices and systems depend on nature and structure of the registering material. Presently in this paper we give

M.K. Zaidi and I. Mustafaev (eds.), Radiation Safety Problems in the Caspian Region, 205-209.
© 2004 *Kluwer Academic Publishers. Printed in the Netherlands.*

great attention to materials, which have potential for application in dosimetry where small doses of radiation are measured.

The fundamental concepts and applications of electret dosimetry have outlined many advantages. The main advantages of electret dosimetry are: excellent linearity; adaptation to all kind of needs; the dose measurement does not depend on the reliability on electronic components present in the dosimeters, dose can be read electrically without destroying the information carried by the electret material. All these indicated features confirm and support the use of polymer and composite materials as registration material of irradiation [8].

EXPERIMENTAL RESULTS AND DISCUSSION:

It is shown in literature, that the values of electret charges σ_E of films of Teflon and polystyrene decreases exponential under action of σ-irradiation [9-11]. These electrets with sensitivity of 1 mrad have been obtained by thermal polarization at voltage of electric field 5 kV/cm and temperature 70° C and had low radiation sensitivity. At action of σ-irradiation up D = 10 kGy (1 Mrad) the values of σ_E decrease from 1.8 10^{-5} C/m^2 to 0.66 10^{-5} C/m^2 [12]. The character of decreasing of value of electret charges depends on radiation dose for PETF, PTFE and PS films at range of dose 1 kGy to 10 MGy is given by equation:

$$\sigma_D / \sigma_E = Const - \lg D$$

This equation is used for dosimetry of photon and neutron radiation [9-11]. The electret films which are pre-irradiated with σ-rays may be used as dosimetric material and at field of doses 6-3.6 10^5 Gy [13,15], the calculation formula has a form:

$$D = 106 - 20 \cdot \sigma \cdot 10^4, \text{ where } \sigma \text{ in } 10^{-4} C/m^2, \text{ and D in Gy.}$$

In case of electrets of PETF at doses up to 2.5 x 10^4 Gy the calculation formula has a form:

$$D = 5 - 0.32 \, \sigma \, 10^4$$

At doses up to 10^5 Gy the linearity dependence between lg D and relatively changing of charges σ is observed for electrets on base of PE and piezoceramics filler PZT-19 [16].

The analyses of structural and conformational changes in polarized films at σ-irradiation show, that in order to use these films as registered material, it is necessary to create in it stable polarized structure and time-stable electret charges and internal electric field. In some cases, at doses of 0.5 - 1 MGy in Polyvinylidene Fluoride (PVDF) on IR absorption spectrum (442, 510, 840, 1280 cm^{-1} lines), the internal structure transformations are observed. It is established, that these transformations depend on crystallization conditions, thickness of the film, degree of orientation of macromolecules, molecular and supermolecular structure, and values of absorbed doses D [17,18].

For dosimetric purposes the electret state of films is created by Corona discharge with electrodes of type of needle-plate at room temperature. The value of polarizing field changes from 4 to 8 kV, and charge time from 5 to 10 minutes. The electret difference potential U_C is measured by method of vibrating electrodes [10] and effective value of surface charge density is determined by equation:

$$\sigma_E = \varepsilon\varepsilon_0 U_C/h$$

where ε_0 dielectric constant, ε dielectric permitivity of electret, h thickness of electret. The value of σ_E can be measured before and after of action of γ-irradiation and on obtained data the calibration curve σ (D) can be built. Then on this calibration curve the adsorbed dose D can be determined. In Figure 1 the characteristic dependence of σ (D) for films of copolymer of tetrafluoriedethylene with hexafluoridpropylene ((σCF_2-$CF_2)_n$- (CF_2- $CF(CF_3))_m$) is lead.

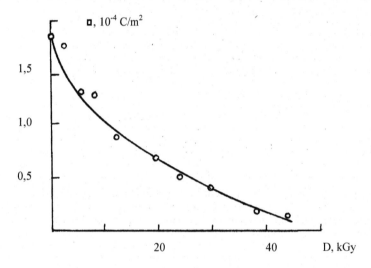

Figure 1. The dependence of charge effective density σ on γ-irradiation dose D for films of tetrafluoridethylene with hexafluoridpropylene ((σCF_2- $CF_2)_n$- (CF_2- $CF(CF_3))_m$).

For modification of properties of registration material on base of polymer films, it is applied by us the method of co-precipitating of ceramic fillers with sizes d < 5 μm [19]. With decreasing of particle size, the surface specific square increases at same volume content of fillers and this leads to increasing of interphase interactions between polymer matrix and ceramic fillers. Therefore, we have the possibilities for modification of ferroelectric properties, electret charges, physical-mechanical properties of materials. In Figure 2 is given the Scanning Electron Microscope (SEM) microphotography of composite material of PVDF + PZT-19 with particles of sizes d < 5 μm.

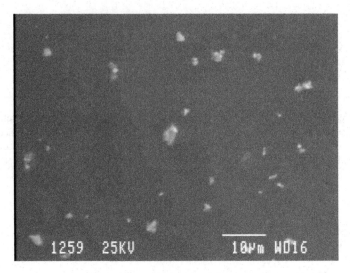

Figure 2. Scanning Electron Microscope (SEM) microphotography of composite material of PVDF + PZT-19 with particles of size d < 5 μm at volume content of fillers of 5 %.

CONCLUSION:

The sedimentation method allows the selection of particle sizes and hence, regulate physical properties of composite material used as registration material for ▫-irradiation and lead to the application of these new composite materials on base of thin polymer films and ceramic particles with sizes about 1 micro-m not only as dosimetric indicators but also as sensors in medical, electronics and biology fields.

REFERENCES:

1. Pikaev A.K. Dosimetry in Radiation Chemistry. M. Nauka (1975). 312.
2. Presentations - The Industrial Appl. of Irradiation Technology, Baku. (2003). 54-64.
3. Andreeva E.A., Torubarov F.S, Chutorskaya O.E., Smirnova S.N. Journal Phiziologiya Cheloveka. (1990).16 (6). 135-141.
4. Lushik Ch.B, Lushik A.Ch. Electron Exicitation Decay with Forming of Defects in Solid State. M. Nauka. (1989). 264.
5. Barsukov O.A, Barsukov K.A. Radiation Ecology, M. Nauchniy Mir. (2003). 253.
6. Bayramov A.A, Gadjiev N.D, Pashayev A.M. Thesis Reports - Modern Problems of Applied Physics and Chemistry. Baku. Scentific-Ecology Company (Azerecolab). (1999) 22.
7. Korobkin V.N, Peredelskiy Y.V. Ekologiya, Rostov na Donu, "Feniks". (2003). 576.

8. Dreyfus G., Lewiner J., Perino D., Wittler W. Electret dozimetry. In 49th Intern. Radiation Protection Affocentr.Execent.Council Meeting Reports. (2003).153-156.
9. Magerramov A. M. Structural and Radiation Modification of Electret, Piezoelectret Properties of Polymer Composites. Baku, Elm. (2001). 327.
10. Lusheykin G.A. Polymer Electrets. M.Chimiya. (1984). 184.
11. Fabel G.W., Henish H.K. (1971). Phys. Stat. Sol. A. 6. 535-541.
12. Zheng W, Kobayashi Y, Hirata K. (1998). Radiat.Phys.Chem. 51 (3). 269-272.
13. Electrets by Red. Sessler. M. Mir. (1983). 342-356
14. Magerramov A.M. (2003). Abst. Conf. Radiation Safety of Caspian Region, Baku. 84.
15. Magerramov A. M. (2003). Chimicheskaya Fizika. 22 (3). 80-83
16. Tutnev A.P, Vannikov A.V, Mingaleev G.S, Saenko V.S. (1985). Electric Effects at Irradiation of Polymers. M. Energoizdat.176.
17. Magerramov A.M, Izvestiya A.N. (1999). Azerbaijan 19 (6). 171-176.
18. Kochervinskiy V.V, Lokshin B.V. (1999). Visokomolek. Soed. A. 41 (8). 1290-1301
19. Hamidov E.M, Tanem, Byorn, Magerramov A.M, Nuriyev M.A. (2004). III Russian Kargin Conference " Polymers-2004". Moscow (in print).

ENVIRONMENTAL IMPACT OF ELECTROMAGNETIC FIELDS

EDWIN MANTIPLY
U. S. Federal Communications Commission
Washington, DC, USA.

Corresponding author: emantipl@fcc.gov

ABSTRACT:

The biological effect of about 100 kHz radio frequency is heating; at lower frequencies the primary effect is electric shock or neurostimulation. Many low-level cellular and animal effects are not understood and difficult to repeat. Epidemiological results of most significance correlate power frequency magnetic field exposure with childhood leukemia. Standards to limit exposure require fairly complex but well understood dosimetric calculations of the coupling between unperturbed exposure fields and the human body. The possibility that brain tumors are correlated with handheld wireless phone use is the subject of current research.

Keywords: EMF, bioelectromagnetics, environmental fields and RF safety.

INTRODUCTION:

The intent of this paper is to serve as an initial resource locator and to introduce current scientific issues to basic researchers new to the safety, risk, and environmental impact of electric and magnetic field exposure. The subject includes both radio frequency (RF) and power frequency (PF) fields. There are many books that cover this broad field; one of the more comprehensive is "Handbook of Biological Effects of Electromagnetic Fields" 2nd edition 1996, edited by Polk and Postow.

A biophysical perspective points toward scientific questions that remain unanswered and the need for a serious long-term interdisciplinary team initiative to gain a depth of understanding of the mechanisms of interaction of weak fields and biological molecular systems for both medical and safety applications. But, research support in the field has not been consistent and has lead to a large uneven literature with limited follow up for interesting results. Many experimental results are presented as simple observations with the standard caveat that the mechanism for the effect is not understood. Efforts to build on an observation; to repeat, extend, and hypothesize; are typically truncated because of an inability to even reproduce an existing observation. It seems possible that weak field effects are artifactual (statistically or methodically) and theoretical models have been advanced to support this position. If the artifactual premise is true then there is essentially no environmental impact due to anthropogenic electric and magnetic fields and further research will presumably waste resources. However, more work is needed even in this case to show that weak effects are artifactual. Alternatively, these effects may be so subtle and unstable that experimental variables that are normally uncontrolled and unimportant

M.K. Zaidi and I. Mustafaev (eds.), Radiation Safety Problems in the Caspian Region, 211-216.
© 2004 *Kluwer Academic Publishers. Printed in the Netherlands.*

become significant. Subtle effects leading to small risks or minor complaints may be important from a public health standpoint considering the ubiquity of exposure to weak fields.

The above mentioned facts should not imply that there are not well-understood acute and harmful effects of fields, however, these thermal and shock effects occur at such high field strengths that they are rarely relevant in the general environment. These effects are universally accepted and used for standard setting not only because of repeatability of experiments but also because they must occur in the biophysical sense. Depolarization of a neuron will cause an action potential and deposition of energy will cause temperature to rise. However, in the same sense, spatially and temporally coherent oscillating electric fields in tissue will cause coherent, but possibly insignificant, motions in ions, dipoles, and zwitterions. This coherent motion contrasts with stronger random thermal motion but might be expected to shift critical equilibrium points for electrostatically driven binding processes. Once covalent bonds are established the molecular machinery of the cell may be modeled as classical electrostatically interacting soft objects of various shapes, flexibilities, and electric charge distributions. Classical molecular dynamics simulation using computers to model biological processes may benefit this field by reducing the gap between simple biophysical models and complex biological systems.

Depending on the ultimate results of a science under construction, the environmental impact of electromagnetic fields may be large or insignificant. On entry into this field of study, reactions range from dismissal to alarm, depending on how critically and selectively the literature is read. Professionals in the field need to repeatedly review the objectivity and supportability of their positions and avoid the polarizing influence of emotional appeals and institutional bias.

THERMAL/NEURAL EFFECTS AND PREVENTIVE EXPOSURE LIMITS

A small number of critical studies form the basis of exposure standards used to prevent neural stimulation or thermal injury. For the RF case at frequencies between about 100 kHz and 10 GHz whole body exposures to strong fields causes a thermoregulatory response that maintains body temperature in the face of significant additional thermal load due to power absorption from the field. Experiments with primates reliably show that at a load or specific absorption rate (SAR) of about 1 to 4 watts per kilogram (W/kg) of body mass, the animals will stop working for rewards. This threshold for whole body thermal overload is consistent across species. While power absorption in the body is not uniform, the temperature remains relatively constant throughout the body because of efficient heat transport by blood flow. Using a convenient human body mass of 75 kg or 165 pounds and an SAR of 4 W/kg suggests an adult human thermal overload power of about 300 watts. Most western occupational limits are set using an arbitrary safety factor of 10 to limit the load to 30 watts and public limits include an additional safety factor of 5 to limit the load to 6 watts. The 6 W load for an adult member of the public is allowed on a continuous basis. Field strength and power density limits are based on the effective absorption area of the body, which at wavelengths close to body dimensions is about the same as the physical area. For example, at 900 MHz the body effective area is about 1 square meter (m^2) so exposure to a wave with a power density 6 watts per square meter (W/m^2) will result in 6 W of absorbed power and in fact the whole body public exposure limit at 900 MHz used in the U.S. is 6 W/m^2 or 0.6 mW/cm^2. Because of active thermoregulation, body temperature may not rise very much during significant loading, so SAR is a more appropriate measure than actual temperature. However, in the case of localized absorption, allowable temperature rise can be used more directly and local SAR limits may be based on organs with poorest thermoregulation or greatest sensitivity to temperature change.

At frequencies below about 100 kHz the body is more sensitive to shock and neural stimulation than heating, so that field exposure limits are based on limiting current density inside the body. The stimulation threshold current density depends strongly on frequency with neurons becoming more sensitive as the frequency is lowered. However, since low frequency external electric and magnetic fields generate currents in the body that are proportional to frequency, the biological sensitivity and coupling tend to cancel and the exposure limits tend to be flat in field strength as a function of frequency.

WEAK FIELD EFFECTS AND PRECAUTIONARY APPROACHES

The particular studies mentioned here are those that are most visible, controversial, or contain information that seems suggestive as to mechanism under the assumption that the reported effects are not due to artifact. Speculation relevant to developing new hypotheses for simulation or experiment is not avoided.

HUMAN STUDIES:

It is understandable to consider human studies and epidemiological literature as the proper source of information about human health risk, yet these studies may suffer from confounding or issues of cause and effect versus correlation where a presumptive but untrue cause is correlated to a true cause. The lack of knowledge of real human exposure and subject biases can either obscure a real effect or suggest an effect that does not exist. Without theoretical and laboratory support it is difficult to consider epidemiological studies conclusive.

Most current epidemiology is focused on cell phone exposure and while there are results showing a tendency for tumors to develop on the side of the head where the phone was used; the general sense is that there is no obvious correlation in the short term that phones have been in use and that continued surveillance with improved exposure assessment is appropriate. Phone exposure assessment is complicated by the variable power and modes of current devices.

The most mature area of epidemiological research deals with residential exposure to power frequency magnetic fields and childhood leukemia. After many studies and more than one independent meta-analysis the International Agency for Research on Cancer [www.iarc.fr] has classified these fields as a possible human carcinogen. The World Health Organization [www.who.int/peh-emf/en] has been working on precautionary approaches to dealing with such a possible risk. In practice, new power line construction has been affected where significant reductions in field exposure can be achieved with relatively low cost changes in line design.

ANIMAL STUDIES:

Assuming that weak non-ionizing fields cause a small increase in human disease risk because of marginal non-specific modifications to physiology and cell signaling (stress) without tissue damage or mutation; animal studies of reasonable size must either detect small physiological effects or somehow amplify the risk. Adaptation may mask a weak physiological change so that looking at the adaptation response itself is an option. Genetic manipulation can be used to produce animals with unusually high disease risk and sensitivity to exposure. Hazard identification studies of normal sensitivity would be expected to exhibit apparently conflicting and inclusive results because the effect magnitude is near the detection threshold or in the noise. Positive controls are generally necessary to demonstrate that an experimental system has adequate sensitivity.

An Australian study of general interest used lymphoma prone transgenic mice in an effort to detect a weak effect of cell phone exposure. About twice as many mice in the exposed group developed lymphoma. The most common criticism of the study was that several mice were free to move about in each cage. This movement and variable coupling between mice results in a wide range of absorbed power in the animals. A more recent follow up study using restrained transgenic mice to better quantify absorbed power did not show an increase in lymphoma. It has been criticized because of a loss of sensitivity due to the high level of lymphoma in the control group possibly due to the stress of restraint.

The early chick embryo has been used to show a protective action of exposure to electromagnetic fields. Apparently, the fields stimulate a protective stress response that is not directly detected. Effectively exposed embryos are less susceptible to asphyxiation. Effective exposure depends on several suggestive field characteristics. In general, sinusoidal fields applied for an adequate time are effective but noisy or short exposures are not effective. Similar patterns of effective exposure are seen for activation of the enzyme ornithine decarboxylase [1].

CELLULAR AND BIOPHYSICAL STUDIES:

A major area of research evolved from the idea that low frequency fields in the range of about 3 to 30 hertz might affect brain-wave activity in the same frequency range. Studies in the 1970's [2] showed that weak ELF fields affected the binding of calcium ions to chick brain tissue. In addition, RF fields amplitude modulated at the same ELF frequencies caused similar effects – suggesting some biological demodulation mechanism [3] that has not been demonstrated except in plant cells at frequencies much lower than those used in the calcium experiments. Effects on the blood-brain-barrier [4] and possible effects on the eye-blood barrier suggest modification of the extracellular calcium binding equilibrium as a possible cause. Tight junctions between capillary cells form the seal for these barrier systems. Tight junctions are formed by cadherin proteins and calcium ions.

A second broad area of research deals with the effect of fields on bone and soft tissue repair [5]. Moderate strength pulsed magnetic fields at low frequencies are apparently effective in the treatment of bone fractures that fail to heal. This area of research began with the observation that electric potentials are generated when bone is stressed mechanically. This "streaming potential" is due to fluid motion displacement of free ions near fixed oppositely charged ions. Pulsed magnetic fields are used as a non-invasive method to generate induced electric fields in the tissue.

The idea that positive ions (calcium) are relatively free to move in solution in the extracellular matrix and that negative organic counter-ions are fixed to the outside of cells (ionized sugar chains) leads to some interesting speculation although quantitative models are lacking. This extracellular "glycocalyx" is a region of concentrated current in tissue at low frequencies where the cell membrane is a good insulator. Oppositely charged ions experience opposing oscillating forces in an AC field. The negative charges at the distal end of polysaccharide chains have the freedom to "wave" back and forth in the field while the positive calcium ions that are relatively free to begin with may efflux because of this motion. A steady magnetic field would impart some rotational component to the motion.

Since the efflux of calcium and other effects are exhibited at specific resonant frequencies and these frequencies in turn depend on the local steady magnetic field, classical cyclotron resonance and quantum mechanical Zeeman splitting theories have been proposed to explain these effects. However, "none has been shown to explain, in an unambiguous manner, the experimentally observed effects in biological systems" [6].

EXPOSURE ASSESSMENT:

In contrast to the study of weak field biological effects, exposure assessment is a relatively straight forward electrical engineering exercise where computer models and measurements can come together to solve the occasional oversight. Even so, the relative ease of measurement makes inexpensive instruments available that are often poorly calibrated. Field calculation has the advantage of being inexpensive, fast, and able to determine fields at many locations. Measurement is necessary where accuracy is important at relatively few locations.

Field Calculation

A radiofrequency field calculation usually begins by applying simple point source (spherical), line source (cylindrical), or plane source (aperture) equations where the radiated power (with various corrections) is divided by an appropriate physical area to give power density or intensity of the electromagnetic wave. These formulas are given in the U. S. Federal Communications Commission, Office of Eng. & Tech. bulletin 65 [http://www.fcc.gov/oet/rfsafety]. At power frequencies and low radiofrequencies in the reactive near field ampere's law is often useful for calculating a quasi-static magnetic field, however, simple quasi-static electric field calculation depends on electrostatic models such as parallel plates which are rarely approximated by real field sources. Software packages are available for general solutions to complex problems but are often expensive or hard to use.

Field Measurement

Where fields approach exposure limits and where reassurance is needed, measurements are feasible and definitive. Most measurements are made with calibrated broadband survey meters, see, http://www.narda-sts.com, http://www.holadayinc.com and http://www.ifi.com for professional commercial instruments. If the intensity due to many sources operating at the same time or very high sensitivity is required a calibrated antenna and computer controlled spectrum analyzer are typically used but here more training is necessary.

Specific Absorption Rate (SAR)

Since RF field limits derived from SAR assume uniform exposure to the whole body, it is common for localized fields to exceed the whole body limit but for the SAR values to be compliant with local and whole body SAR limits. This is typically the case for transmitters held near the body such as cell phones. Here, laboratory measurement of local SAR has been standardized and is complemented by computer simulations using Finite-Difference-Time-Domain (FDTD) calculations.

CONCLUSIONS:

Many low-level cellular and animal effects are not understood and difficult to repeat. Epidemiological results of most significance correlate power frequency magnetic field exposure with childhood leukemia. The possibility that brain tumors are correlated with handheld wireless phone use is the subject of current research.

Acknowledgement: The views expressed are those of the author and do not necessarily reflect the views of the U. S. Federal Communications Commission or the U. S. Government. NATO provided full financial support to present this paper at the NATO Advanced Research Workshop, Radiation Safety Problems in the Caspian Region, held in Baku, Azerbaijan during September 11-14, 2003

REFERENCES:

1. Litovitz, T. A., Drause, D., Penafiel, M., Elson, E. C., and Mullins, J. M. (1993). The role of coherence time in the effect of microwaves on ornithine decarboxylase activity. Bioelectromagnetics 14, 395.
2. Polk, C. and Postow, E. (1996). Biological Effects of Electromagnetic Fields (BEEF). Boca Raton. CRC Press. 546-550.
3. Polk, C. and Postow, E. (1996). BEEF. Boca Raton. CRC Press. 13.
4. Polk, C. and Postow, E. (1996). BEEF. Boca Raton. CRC Press. 471 and 507.
5. Polk, C. and Postow, E. (1996). BEEF Boca Raton. CRC Press. Chapter 5.
6. Polk, C. and Postow, E. (1996. BEEF. Boca Raton. CRC Press. 560.

ECOLOGICAL PROBLEMS OF POWER ENGINEERING – ELECTROMAGNETIC COMPATIBILITY AND ELECTROMAGNETIC SAFETY

M.S. QASIMZADE, F.S. AYDAYEV, V.M. SALAHOV
Azerbaijan Scientific Research Institute of Power Engineering and
Power Designing, National Academy of Sciences
Baku, AZERBAIJAN

Corresponding author: vilayet79@yahoo.com

ABSTRACT:

Man-caused electromagnetic fields affect dangerously to the health of the people, population for its specific significance. The professional staff being forced to work at the high and wide frequency spectrum of electromagnetic fields for a long time. Active electromagnetic safety from radiations of sanitary-epidemiological standards are being analyzed. The results of numerous full-scale study of electromagnetic fields of power frequency in the Azerbaijan high voltage electric power system are being presented. Summary of real state of electromagnetic safety proposing concrete protective actions is presented.

Keywords: electromagnetic field, compatibility and safety, biological action, measurement, control, monitor, and instrumentation.

INTRODUCTION:

All problems of electromagnetic compatibility and safety in the field of engineering and biological research, its urgency in current stage of technical progress, wide electrification and electronics in the field of life and activity are under consideration. All types of technical erections working at electron power radiate man-caused electromagnetic field exerting essentially negative influence to surrounding physical and biological objects, human beings health. Power industry is an actual problem. Several data carrying out at energy and power projecting research institute in the field of electronic safety, measurement and monitoring of electromagnetic fields at the industrial frequency, full-scale measurement of electromagnetic field at the high voltage establishments of Azerbaijan electro power systems, personal electromagnetic safety measures were conducted and discussed.

Power engineering, playing a great role in the contemporary economy and social life of society is also one of the unfavorable fields of production from the ecological point of view. All the processes focused on the electricity production, transmission and consumption are accompanied by the damage to the environment, in particular at the traditional technology.

M.K. Zaidi and I. Mustafaev (eds.), Radiation Safety Problems in the Caspian Region, 217-223.
© 2004 *Kluwer Academic Publishers. Printed in the Netherlands.*

A new aspect of ecological problem arousing in power engineering has caused special alarm of scientists and specialists recently that concludes with the electromagnetic compatibility and electromagnetic safety both in technical and biological spheres.

All the equipment including electric power, electro–technical, radio-electronic, computer installations, operated with the utilization of electric power produces electromagnetic fields (EMF) of wide spectrum frequency and different intensity.

Being imposed on natural EMF in the biosphere, these fields have essentially negative impact on surrounding physical and biological objects, violating their normal way of functioning [1].

Nowadays, the rapid development of electrification is observed in our lives and social activities and problems caused by these took a biter character and need to conduct further studies to find their effects on ecology, medical, biological, technical, social and other aspects of human life.

BIOLOGICAL ACTION:

The human organism is considered as the bioenergetics system. The mechanism of EMF biological impact on the living organism depends on all key parameters of field (the form of curve, spectral composition, intensity, frequency and so on), but the consequence depends on the absorbed dose of the electromagnetic power. EMF of industrial frequency concerns to the non ionizing radiations. The frequency, close and equal biorhythms of organism have more substantial impact (on the human brain it is ranged at 5-13 Hz). The important parts of organism such as nervous, immune, sexual, cardiovascular and others are the most sensitive toward the impact of EMF. Irradiation with the nonionization radiations (gamma, X-ray and so on.) of $f = 10^{16} - 10^{24}$ Hz brings about collapse of skin structure, formation of new complexes with the corresponding peculiarities, devastation of large amount of living cells, destructed from their plasma membrane.

The impact of high and ultrahigh frequency microwaves and radio waves ($f = 10^4 - 10^{10}$ Hz) is accompanied mainly by the thermal effects, heating of living skin that brings about morphological and functional damage to the human organism.

As to the low frequency nonionizing EMF (including EMF of industrial frequency), according to the advanced considerations, the voltage and the current leaking through the human body (current density) have main biological impact, stipulated by the charges induced under the influence of external electric and/or magnetic fields and their coincidence with the own current and voltage of organism. In the different countries, the accepted admissible value of current leaking through the human body differs and is ranged at 5-20 mA.

The admissible value of current of industrial frequency is accepted as 10 mA in our country. The current less than 1 mA (current density $j \square 0.1$ mcA/см^2), as a rule is not felt by a human being. Convulsive reduction of muscles can bring about cessation of respiratory movement at 25-40 mA. At the density range $j > 100$ mcA/cm^2, violation can be observed in the working of the heart and may even halt the heart activity. It is appropriate to utilize electric current (current density) leaking through the human body from the point of view of biophysical substance [2,3].

But, measurement of electric and magnetic fields tenseness value is more appropriate for the safety of human beings.

Unfortunately, the accepted unique world sanitary-hygienic admissible standard for the tenseness of electric (E) and magnetic (H) fields lack currently, and the task on standard identification expects solution of their decision.

World Organization for Health Protection (WOHP) in association with the International Commission on Radiological Protection (ICRP) recommended the following norms of stay in the electric and magnetic fields of industrial frequency:

Electric fields

For population
under E ◘ 5 kV/m –time of stay is not restricted;
under E ◘ 10 kV/m – time of stay is no more than 3 hours a day;
For the specialists
under E ◘ 10 kV/m – time of stay is the whole working day;
under E ◘ 30 kV/m – time of stay is no more than 3 hours a day.

Magnetic field

For population
under H ◘ 80 A/m (B◘0,1 mTl) – time of stay is not restricted;
under H ◘ 800 A/m (B◘1 mTl) – time of stay is no more that 2 hours a day;
For specialists
under H ◘ 600 A/m (B◘0,5 mTl) – time of stay is the whole working day;
under H ◘ 4000 A/m (B◘5 mTl) – time of stay is no more than 2 hours for the working day.

Basing on the accumulated scientific data for today, intensity of magnetic field of industrial frequency with the value of magnetic induction exceeded 0.3-0.4 mcTl on condition of chronological impact is the carcinogen of environment [5].

Sanitary-epidemiological rules and standard introduced in Azerbaijan for the residential buildings and apartments recommended the density of magnetic flow (induction) of industrial frequency as B ◘ 10 mcTl that according to the specialists it is not justified scientifically and overestimated.

Safe level of MF of industrial frequency intensity equal to B = 0.2 mcTl is recommended by the national organizations of countries like Sweden, USA and others, as threshold while installing correlation between B and leukemia under continuous stay. Standard on the electric field have been accepted in compliance with the instructions for conducting work places under the labor law of Azerbaijan. These standard is:

Standard time for human being stay in the magnetic field during a day.

Tenseness EF, kV/m	Admissible time of stay
Less than 5	Not restricted
From 5 to 10	No more than 3 hours
From 10 to 15	No more than 1.5 hours
From 15 to 20	No more than 10 minutes
From 20 and above	No more than 5 minutes

Note: In case the tenseness of EF exceeds 25 kV/m, the work should be conducted by the application of protective means.

With regard to the standards on MF of industrial frequency, above-mentioned standard of WOHP is used as a guide in the Republic.

Under the influence on the human being, EMF of industrial frequency can cause violation such as indisposition up to heavy disease-violation of central nerve and heart-vascular systems, headache, dizzy spell, dream violation, depression, blood composition change and so on. Guideline exists on the direct field impact on the exchangeable process at the cell level, on the threat of arousing genetic violation, cancer and nerve diseases.

It is supposed that medical consequences, such as cancer diseases, change in behavior, loss of memory, Parkinson's and Alchmayer's diseases, AIDS, the syndrome of sudden death of a healthy child and the growth in the number of suicide are the results of EMF impact.

The electricians suffer brain cancer 13-fold more than the people of other specialties. The fact of swelling rise in the children under the irradiation by the magnetic field 60 Hz is totaled 2-3 mGauss (0.2-0.3 mcTl) during the several days, even hours. Such fields are radiated by the TV-sets and computer displays.

The measurements on earth magnetism were conducted among 230000 workers of Saint-Petersburg electrified railway system had shown that the foot-platemen of electric locomotive, the working places of which are disposed next to the electric motors suffer from the hypotonia, ischemic disease of heart twice frequently than their colleagues, working at other parts of the electric engine in the same type of electric locomotives.

The result data of epidemiological research conducted in monitoring 50000 residents of Sweden, lived at different times at a distance of 800 m from 200 kV and 400 kV TL, showed substantially direct interrelation of EMF of industrial frequency for the development of cancer, in particular children leukemia. At the same time, stay in the sites of powerful electromagnetic radiation increases the risk of blood cancer almost fourfold. The literature shows that cancer of lungs happens 20-fold more for the smoking people at the electromagnetic fields.

Interaction of EMF with the other harmful physical factors strengthens their destructive impact. These and majority of other research, conducted in the last decade in the different countries of world confirm the danger experienced with electromagnetic fields. It is threatening the lives of the people living around transmission lines. Electric power has not only favorable impact. The problem is especially attractive for the personnel of power engineering system with the installations of high voltage and supervoltage, capacious electro-technical facilities and others, depending on their professional activity of forcedly long stay in the fields of industrial frequency and EMF of high intensities. The problem is for the population who are subjected to the long-run nonprofessional irradiation from the facilities of low voltage, residential electrotechnical apparatus, devices, and utility grids.

Much attention is paid to the problem by the national and international scientific-technical organizations. European Bioelectromagnetic Association, International Commission on the Protection from Nonionizing Radiations and Research Laboratory Military Air Forces USA are existed to this favor. The measures and means of protection taken against the harmful impact of EMF, it is necessary to obtain valid information on the EMF features and have corresponding means of measurement and control. The impact of EMF can be decreased by the following means:

1. Increasing the distance between the sources and object of irradiation. (Protection by distance)
2. Decreasing the stay duration in the site of intense EMF (Protection by time).
3. Screening the EMF sources (Protection by screening)
4. Screening the irradiation object.
5. Decreasing the emission intensity of EMF impact on the voltage, current force and other design-technical decision (transpolation of phases).
6. Installation and fencing of dangerous zones, identification of isolation sites (TLs) and so on.

Nowadays, this or that means of protection are applied depending on the situation aroused. At this time, the difficulty of protection from the magnetic fields in comparison with the electric fields should be noted specially.

MEASUREMENT OF UNIONIZED ELECTROMAGNETIC FIELDS:

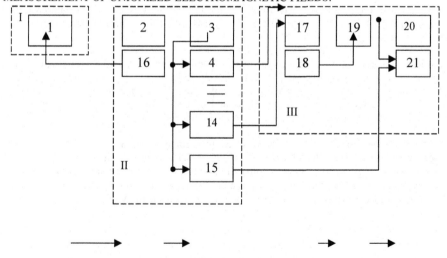

I – Remote with transmitter Holl;

II –Unit for analogue processing of composition.
2 – Booster; 3 – Rectifier; 4 – 15 – Comparator; 16 – Generator of supply for transmitter Holl.

III – Unit of digital processing and indication of composition.
17 – Constant reminding facility; 18 – Generator of support frequency;
19 – Programming meter; 20 – Indicator; 21-Facility of light and sound signalization.

One of the devices used is the dosimeter of magnetic field, and the structural scheme of which is provided above. Some research work was conducted in Azerbaijan on the electromagnetic safety directed to the creation of measuring means, controls, dosimeters of EMF of industrial frequency, and also on the provision of electromagnetic safety of Azerbaijan Power System service personnel.

A 3-component voltmeter of electric field, 3-component voltmeter of magnetic field, dosimeter of electric and magnetic fields, all for industrial frequency measurements have been developed locally and being used.

The scheme allows to determine the cumulative current time of stay (Ts) of personnel in the magnetic field of various intensity industrial frequency (Bi, Hi) for 8 working hours a day and in case of increase of this time for the standard indications, to provide the light and sound signalization. Cumulative time of stay Ts is calculated using equation posted below:

$$Ts = 8 \ (Tst1/Tst2 \ /+Tst2/tgh2+....+Tsti/tsti \ )$$

in which Tst1, Tst2, …. Tsti – actual time of stay in the site of magnetic induction (with the tenseness) $B_1(H_1)$, $B_2(H_2)$, ……Bi(Hi) accordingly, tst1, tst2, tsti – standard (admissible) time of stay value at corresponding values Bi(Hi).

Considering both scientific and practical values, natural research of electromagnetic fields in the Azerenerji Power System substations of 500/330/220/110 kV [4], the measurement of electromagnetic fields has been conducted directly under TL; measurements of EMF in some substations were reconducted in spring-winter and spring–summer periods with the purpose of exposing the impact of seasonal factors such as environmental changes.

The maps of electric and magnetic tenseness have been complied for all the objects as a result of research conducted. The value of tenseness of electric fields in the sites of frequent visits by the personnel for substations with various voltage indoor and outdoor switchyards contains outdoor switchyard 500 kV – E ◻ 18 kV/m; 330 kV E ◻ 16 кV/m; outdoor switchyard, indoor switchyard 220 кV E ◻ 14 kV/m; outdoor switchyard, indoor switchyard E ◻ 10 kV/m.

Beyond the limits of substation at the distance up to 1 km under the high-voltage Tls E ▫ 1 kV/m. The tenseness of magnetic field measured at the substation "Sangachal" and "Agsu" happened to be at the range of 20 A/m, i. e. within the standards.

The value of E substantially exceeds long–run admissible sanitary-hygienic standards and testifies the unfavorable standard of electromagnetic safety in the power system and the necessity of taking protective measures. The substation of Sumgait Power network is under special concern where high value of E has been exposed both in 220 kV indoor switchyard and 110 kV indoor switchyard, and personnel of which are constantly complaining of their health problems. The impact of seasonal factors in the climatic conditions of Azerbaijan occurred to be insignificant, but upgrade of equipment is accompanied by the definite changes (reduction) of E.

CONCLUSION:

The application of bioprotection means (screening of source-grounded grids, screens, and so on) is recommended on the sites of E with high voltage, and also strict restrictions on personnel staying in the dangerous sites should be observed. While complex attesting the working places in compliance with the safety techniques and labor protection rules, installed in the power system, the results of this research are widely used to forecast the behavior of high-voltage equipment and correcting the designing decisions and so on.

REFERENCES:

1. Elektromagnitnie polya v biosfere. (1985). USSR Academy of Sciences, Moscow. Publisher Nauka. 375.
2. Vliyanie elektoustanovok visokogo napryajeniya na okrujayushuyu sredu. (1988). Translations of Reports of the International Conference on power systems. Publisher EnergoAtom. 104.
3. Dyakov A.F, Levchenko I.I, Nikitin O.A. (1997). Elektromagnitnaya obstanovka i vliyanie ee na cheloveka. Elektrichestvo No 5.
4. Qasimzade M.S, Aydayev F.S, Mamedov I.M. (2002). Elektromagnitnie polya i zdorovye cheloveka. Moscow - St.Petersburg. 151-152.
5. III international conference. (2002). Electromagnetic field and human health. Fundamental and applied researches, Moscow.
6. WOHP Bulletin. (2001). No 26. "Electromagnetic fields and public health - Extremely low frequency fields and cancer"

THE PROFESSIONAL-ORIENTED REGIONAL RADIOECOLOGICAL COLLABORATION OF SOUTHERN CAUCASIAN STATES

M. AVTANDILASHVILI, S. PAGAVA, Z. ROBAKIDZE, V. RUSETSKI.

Radiocarbon and Low-Level Counting Section, Nuclear Research Laboratory, Physics Faculty of I.Javakhishvili Tbilisi State University (TSU), Tbilisi, 0128, GEORGIA

Corresponding author: spagava@access.sanet.ge

ABSTRACT:

Civilized Universe aims "To Live and Collaborate into Safe – Ecologically Pure Environment". So Environment Protection is considered as priority of all nations. Citizens of Southern Caucasian Newly Independent States (NIS) realize during years of independence! However, in Georgia a collective nature between officials and representatives of research and public bodies to resolve radioecological problems is noticeable. Therefore, researchers from TSU suggest to discuss establishment of Professional-Oriented Regional Radioecological Collaboration group and support further activities.

Keywords: Radiation, research, cooperation and support.

INTRODUCTION:

The Collaboration aims to study (as Independent Expert Group) the radioecological situation in areas of Southern Caucasus, to assess the risk caused by the influence of ionizing radiation on population, to create broadly accessible regional radioecological database and to assist popularizing of radioecological studies, upgrading Southern Caucasian population's erudition in the field of radioecology and radiation safety and improvement of collaboration between NGOs' and governmental institutions.

The success of the presented collaboration under NATO and other institutions support will create a regional collaboration to solve one of the most timely environment problem and involvement of research bodies from other countries of Caspian Region, as the idea of creation the ecologically pure living space is concordant with interests of Eurasian population.

Today Environment protection is considered as a priority of the national interests - as one of the main component of sustainable development of the states, including NIS of Southern Caucasus and of the whole Eurasian area. The NIS citizens, especially in Southern Caucasus, had received the grave ecological inheritance in the form of degraded ecosystems, polluted by various anthropogenic, including radioactive, pollutants, realized this clearly only during the last years - years of independence. Though it is generally known that during the past 50 years environment was globally impacted more, then during the whole period of civilization's history.

M.K. Zaidi and I. Mustafaev (eds.), Radiation Safety Problems in the Caspian Region, 225-229.
© *2004 Kluwer Academic Publishers. Printed in the Netherlands.*

In Georgia, officials and representatives of research and public institutions is not observed in the field of solving the radioecological problems. In 1996, at the international conference "Decade after Chernobyl: Assessment of the Accident's Consequences", where the existent and potential consequences of the accident were analyzed in the presence of up to 800 experts from 71 country including two officials from Georgia and more then 200 journalists, condition in Georgia and in Southern Caucasus as a whole stayed out of the discussion, because officials did not present necessary information! Therefore, on the map showing pollution of the European territory by [137]Cs Southern Caucasus is presented as a "white spot" (Fig. 1) [1]. Consequently, countries from this region did not receive appropriate aid from the international institutions for rehabilitating the polluted territories and implementation of other purposeful programs.

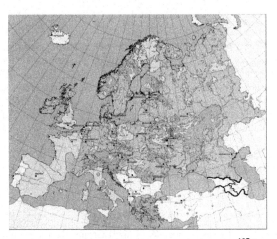

Fig. 1: Pollution of the European territory by [137]Cs.

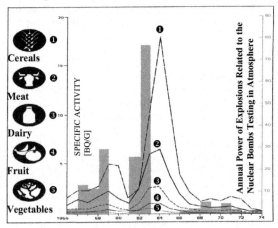

Fig. 2: Variation of ^{137}Cs concentration in foods during 1955-1975.

The greater part of Southern Caucasian population is not aware of the radioecological problems coming from contamination by anthropogenic pollutants – environment and economical activities. Public does not realize the potential risks, caused by the ionizing radiations. They depend on person's lifestyle - behavior, occupation and living place. At the same time, skills of dissemination the information of such type are insufficient.

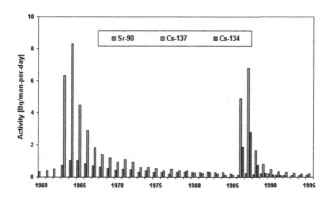

Fig. 3: Annual average concentration of anthropogenic radionuclides in adult daily allowance in Germany

The native population is not informed about pollution due to fallout from testing conducted during 1960-90. Southern Caucasus is contaminated by long-lived anthropogenic radionuclides such as ^{14}C, ^{90}Sr and ^{137}Cs. Consequently, in human's food allowance, an increase was documented [2,3]. (Fig 2,3).

The item of concern is also the circumstances that the safe operation of Armenian nuclear power plant is not discussed openly. The Armenian nuclear power plant first time was set in operation during the Soviet period. Its power reactors (VVER-440/230s) with depreciation period of 25 years were started: first in 1976, second in 1979. In December 1988, after the strong earthquake, the Armenian government decided to stop the nuclear power plants and its full decommissioning was planned. But in 1995, the second reactor of the Armenian nuclear power plant was set in operation again and is still operating in spite of coming to the end of its depreciation period. So it is impossible to exclude the possibility of accident at the Armenian nuclear power plant. The possible scale of the mentioned accident's consequences and their potential impact on Southern Caucasus, and in the Middle East as a whole, can be estimated by taking into account that in VVER-440/230s reactor's core during 3 years operating period approximately $2x10^{19}$ Bq of activity is accumulated, while after the Chernobyl Accident to atmosphere approximately $2x10^{18}$ Bq of activity was released [4].

SUGGESTION:

Therefore, proceeding from the above mentioned risk from nuclear power plants in the neighboring states and taking into consideration the Southern Caucasian realities, it is suggested to establish a Professional-Oriented Regional Radioecological Research and Educational Collaboration Center of Southern Caucasus, as NGO — independent expert group to conduct collaborative research and educational activities such as:

1. To study the radioecological situation in separate areas of Southern Caucasus.
2. To assess the potential risk for population caused by the influence of ionizing radiation.
3. To create broadly accessible regional radioecological database.
4. To assist popularizing of radioecological studies and upgrading Southern Caucasian population's erudition in the field of radioecology and radiation safety;
5. Improvement of collaboration between NGO's and governmental institutions.

Success of the presented Collaboration under NATO or other institutions support will create obvious case of the regional cooperation to solve environmental problems; Preconditions for enlargement the collaboration by involvement research bodies from other countries of Caspian Region, as the idea of creation the ecologically pure living space is concordant with interests of Eurasian population. Collaboration of the similar profile, established by undertaking of authors, operates in Georgia since 1998. This National Radioecological Collaboration was established by active support of Georgian SOROS Foundation — Open Society – Georgia Foundation (OSGF). This Collaboration unites following research bodies:

Radiocarbon and Low-Level Counting Section, University, Nuclear Physics Chair of Physics Faculty, Analytical Chemistry and Eco-chemistry of the I.Javakhishvili Tbilisi State University. Selection and Plants Protection Department, Batumi Botanical Gardens, Toxicology Laboratory of Marine Ecology and Fishery Institute, State Maritime Academy, State Department of Geology and Mining, Batumi State University, Hydrometeoservice, Adjara Autonomous Republic and Non-governmental organization – Radioecological Research and Educational Association Radioecology-21.

The wide-scale scientific interests of different institutions of the collaboration gave possibility of rapid selection of professional-oriented task group in accordance of aims and workplan of running or planned projects. Taking into consideration the statements - "The media have been guilty of exaggerations and distortions, priorities have become badly misaligned, the subject of environmental protection has become deprofessionalized to a disturbing extent, and the rigidity of the positions taken by some environmental activists has many of the characteristics of religious dogma" [5]. The Collaboration published booklets with research results and recommendations from international institutions in the field of environment protection and, particularly, radioecology. During the last years 5 different popular booklets were translated to Georgian. "On Rehabilitation and Protection the Black Sea and Sustainable Use of its Resources" [6] issued by IAEA. Now arrangement of translation of and wide discussion around recommendations formulated in Environmental Performance Review (EPR) of Georgia, issued by United Nations Economic Commission for Europe is planned [7].

CONCLUSION:

Regional Radioecological Center, be created by collective efforts of all concerned to conduct consultation, research and educational collaboration.

REFERENCES:

1. IAEA Bulletin. (1996), 38 (3, 5).
2. Pagava S. (2002). Radioecology. Problems & Prospects. Env.Radiation. No. 5.
3. Radiation Doses, Effects, Risks. (1985).UNEP.
4. Pagava S. (2000). National Radioecological Concept. Env.Radiation, No. 4.
5. Eisenbud M., Gesell T.F. (1997). Environmental Radioactivity. Acad. Press, 4th ed.
6. Pagava S. (1998). A Sea of Changing Fortune. Env.Radiation, No. 3.
7. Environmental Performance Review of Georgia. (2003). UNECE.

PUBLIC OPINION AND ITS INFLUENCE ON PROSPECTS OF DEVELOPMENT OF NUCLEAR POWER.

E. A. RUDNEVA, A. V. RUDNEV, N. P. TARASOVA*
Noncommercial Organization, Young Center
Creative Development Of The Person
str. Grusinsky val, 28/45, Moscow, RUSSIA
* Institute of Chemistry and Problems of Steady Development of Russian Chemical-Technological University, D. I. Mendeleev, Miusskaja sq. 9, Moscow, RUSSIA.

Corresponding author: rudalex@mail.ru

ABSTRACT:

The absence of the authentic, accessible and clear for various layers of a community information about a place of nuclear power in the maintenance of a vital level of the population, and about its influence on an environment and on the person's health frequently determines the negative public opinion formation in the questions of nuclear energy. In turn, this public opinion can form the basis for the acceptance by government before making any strategy on long term. Other competing power generating sources are hydroelectric, thermal power generating plants

Keywords: Nuclear power, hydroelectric, thermal power plants, gas, coal and petroleum.

INTRODUCTION:

The majority of the people living on this planet are anxious only in the destiny of their family and several close friends and the nearest period of time, the few excite a problem for the city or of the country. Others who are in great majority, want to satisfy the requirements immediately and behave on a planet as if the future does not concern them. There are boundless, spend resources, and pollute the environment.

The person plans the activity for any period of time: someone for the same day, some for the week, and very few for the year. The circuit offered by D. Meadows (Table 1) is rather evident. However, as the biosphere has survived completely, other thinking is necessary. The forecasting development of a community even on nearest 25 years is necessary. It does not suffice the need of political leaders and to the persons accepting their decisions.

The priority task in the field of steady development of Russian Federation is the definition optimal of ways and sources of this development. One of the most important factors of the state, determining steady development, is the power. Despite significant success in the field of economy of electric power, the tendency of growth of manufacture and consumption of energy is at present kept. In this connection the various structural variants of the development of power are considered.

M.K. Zaidi and I. Mustafaev (eds.), Radiation Safety Problems in the Caspian Region, 231-239.
© 2004 *Kluwer Academic Publishers. Printed in the Netherlands.*

232

TABLE 1. DISTRIBUTION OF THE PEOPLE ON GROUPS DEPENDING ON PROBLEMS, EXCITING THEM.

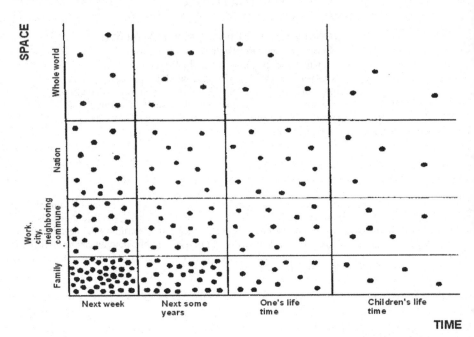

The limited raw power resources define polypower structure of manufacture of energy as optimal. Thus the nuclear power acts as one of the basic components in this structure. The experience of last decades shows, that despite assigned alternative sources of energy of hope (the energy of the sun, sea waves and wind) is not observed while special success of their use in this world. The air turbines are most suitable for the prompt introduction in practice. For example, for maintenance of Great Britain the electric power require the 600 thousands turbines. In view of the area borrowed (occupied) by one turbine, all turbines will borrow 35% of a common surface of Great Britain. It is obvious, that the further development of power based on use of the hydrocarbon raw material (petroleum, coal and natural gas), will restrain by exhaustion of fuel resources.

The analysis of separate traditional and alternative sources of energy shows, that per the nearest decade by the basic sources of reception of energy will stay organic, nuclear fuel and water-power engineering, and it concerns both to the world, and to Russia. From the point of view of security by resources, Russian Federation has an opportunity on a long period to satisfy the needs (requirements) with organic fuel and hydraulic power as far as such approach to a domestic Fuel-Energy Complex (FEC) is justified.

For an answer to this question, the basic parameters of water-power engineering was considered. The power generation from organic fuel and nuclear power effects on such aspects, as: economic, ecological, and also on a degree of risk for the personnel engaged in manufacture and the population, living in immediate proximity from FEC (Table 2,3,4).

Table 2. Type of station for development 33.4 Gigawatt hour of electric power.

Type of station/ Capacity of each of them (Gwatt)	Hydroelectric power station	Thermoelectric power station	Nuclear power station
Quantity of energy carrier for making 33.4 Gwatt hour (120 10^{12} Joule)	Water – 700 10^6 ton	Coal – 6400 т Petroleum – 4600 м^3 Gas – 536000 м^3	Uranium – 1.5 – 2 kg
Aftermath for the personnel (deaths/Gwatt hour)	No findings	Coal – 17.3 Petroleum– 3 Gas – 1	0.7
Prime cost of energy (Cent/kW hour)	No findings	2.5 – 4	2.2 - 3

TABLE 3. POLLUTION ATMOSPHERE OF A VARIOUS TYPES (CAPACITY OF STATIONS – 1 GWATT)

Type of pollutant/Type of fuel	SO_2.1 Ton/24 hours	NO_2.1 Ton/24 hours	Ashes-1 Ton/24 hours	CO_2-1 Ton/24 hours	Radioactive component Activity/Second
Coal (6400т)	380	60	12	1.4	$3.7\ 10^4$
Petroleum (4600 м^3)	14.5	60	2	0.03	$3.7\ 10^2$
Gas (536000 м^3)	0.04	34	1.2	0.02	-
Nuclear Fuel (Uranium 2.5kg)	0	0	0	0	slightly

TABLE 4. THE CHARGE OF NATURAL RESOURCES FOR MANUFACTURE 1 GWATT PER YEAR OF THE ELECTRIC POWER IN COAL AND NUCLEAR FUEL CYCLE.

Resource	A nuclear fuel cycle	A coal fuel cycle
The Earth, hectares	20-60	100-400
Water, million m^3	32 50 - 200 (0.1% uranium in the autunite) 1500 uniflow cooling	21
Materials (without the account of fuel – thousand tons.	16	12
Oxygen Million Tons	-	8

The consequences for the personnel of the enterprises:

The important parameter of social advantages of nuclear power is a professional risk of the workers of this branch. The data for 1996 on risk professional mortality show, that in nuclear power it appreciably is lower, than for power at a coal, petroleum and gas [1].

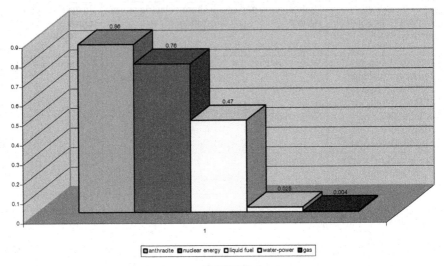

PICTURE -1

According to the data, the conducting factor of an irradiation of the population are natural (65.5 %) and medical (29.5 %) sources of the ionizing radiation [2]. The contribution of industrial sources to a collective dose of an irradiation of the population makes about 1%.

Ecological aspect:

The problem of protection of an environment are considered, as one of the most important global problems of modern age. The objective analysis of activity of nuclear branch in discussed below. The traditional parameter of influence on environment and its estimation of the contribution to industrial parameters allow to assert. The nuclear branch does not concern to number of main sources on one of the basic parameters of pollution. The contribution of this branch on one of the considered parameters does not exceed 5%.

Inevitable emissions of polluting substances in an atmosphere with burning organic fuel, and practically absent in a Nuclear Fuel Cycle (NFC) - bright ecological certificate of nuclear power system (NPS) (Picture 1).

The contribution of nuclear power of Russia is rather insignificant and makes about 14 % from common sources of the electric power and nevertheless results in essentially reduced amount of harmful emissions in the atmosphere that are inevitable with the operation of Thermal Power Stations (TPS).

In conditions of normal mode of operations the activity TPS is accompanied by emissions in an atmosphere of huge amounts of chemical polluting substances, oxides of carbon, nitrogen, sulfur, formaldehyde of dangerous substances having cancerous and mutagenic properties: benzohyrene and number of other heavy metals. The population living in this area of emissions from TPS face ecological problems at local level, the polluted air in cities cause deterioration of health of the population. On a regional and planetary scale, acid rains and "greenhouse effect" add to this. As against NPS, whose emissions are strictly normalized and are subject to the most severe control, the functioning TPS is accompanied by infringement of specifications of emission of one or several substances. Some of the substances found in emissions have very bad biological effects.

The emissions of polluting substances in an atmosphere are connected, basically, to the initial stage NFC - production and enrichment of uranium ores. However greatest contribution give auxiliary TPS and boiler enterprises of branch, on which it shares more than 75 % of emissions of harmful substances. On the other hand, NPS develop large amount of radionuclides, which are absent in TPS.

The burning of the coal in atmosphere considerably add some radionuclides. The air space measurements were carried out on working TPS in Russia and the Ukraine have shown, that the radioactivity of an air pole coal - working TPS exceeds 5 times radioactivity above NPS. The modern activity of nuclear power plants does not overstep the bounds strictly of norms of influence. Radioactive emissions and dumps of the enterprise NFC are much below. Last year, on all NPS emissions on inert radioactive gases and long-lived radionuclides have not exceeded 3%, and the dumps radionuclides made 40-50% from allowable size.

The delivery of fuel from places of production up to the consumer contributes pollution to the air. The transportation of the coal utilizes more than 20% of all rail transportation, thus the pollution closed to a railway bed territory by coal-dust emission from cars is inevitable. The pollution of an environment occurs as other sources also contribute and the outflow of petroleum from oil pipelines and gas from gas pipelines. The scales of pollution are non-comparable to pollution by transportation of Fulfilled Nuclear Fuel (FNF), which does not render practically any adverse influence on an environment.

On data given in the Table 2, 3, 4, it is possible to make a conclusion that the atomic engineering is competitive in comparison with traditional organic fuels on economic parameters and ecological facts. A degree of risk for the personnel engaged in NPS is less than in TPS. Thus, nowadays atomic engineering appears by the most real way of manufacture of energy, which is capable to ensure growing requirements of mankind for energy. However there is a lot of problems, owing to which the public is concerns to this way of manufacture of energy. One of them is the absence of the authentic information about a place of nuclear power in life of a community about influence on an environment and health of the population. Really, as the data of sociological research testify, some layers of a community are badly informed on the basic tendencies of development of a civilization, its problems and ways of their overcoming.

The sociological interrogation of various groups of a public on problems of development of nuclear power were studied and results of public opinion about problems of safety of functioning of atomic engineering are shown in Table 5.

TABLE 5. SPECIFIC PARAMETERS WATER USE OF POWER STATIONS, M3 (MWXH).

POWER STATION TYPE	FRESH WATER VOLUME	VOL. TURN-AROUND AND REPEAT-EDLY USE WATER	VOL. OF IRREV-ERSIBLE LOSSES OF WATER	VOLUME OF MANUFACTURING WATER BE FUSED IN OPEN POOLS			
				TOTAL	INCLUDING		
					BE WASTED	LEGALITY	NON-LEGALITY
TPS	26.3	84.7	4.0	17.5	3.1	1.1	13.3
NPS	51.4	151.3	2.4	48.9	0.1	0.2	48.6

As object of research some categories of the respondents were chosen:

1. The population of Moscow

2. The experts working in sphere of chemistry

3. The teacher of high schools.

The results of interrogation of the population have shown, that the population considers atomic power stations as the most dangerous way of manufacture of the electric power. In opinion of the population, the power future of Russia should be Hydroelectric Power Station (HPS) as illustrated in Picture 2 and 3.

Thus it is possible to note, that nowadays in a power balance of Russia the alternative sources borrow < 1 % from all manufacture of energy. One of the reasons, which can explain negative public opinion on nuclear power, is the unsufficient knowledge of a community on the given question. The results of interrogation of the experts have shown, that from three ways, submitted for an estimation, of manufacture of energy by most dangerous are recognized TPS. The majority of the experts have estimated NPS as "rather dangerous". The special importance was given to have the results of interrogation from the teachers (Pictures 4 and 5)

With comparison of three most widespread ways of manufacture of the electric power by most dangerous are recognized NPS and the most safe - HPS. In a community is observed the negative estimation of a degree of danger of NPS. While the experts consider, TPS as most dangerous.

Picture 2. Indexes of danger for an environment of the enterprises which produces an electric energy (opinion of the population).

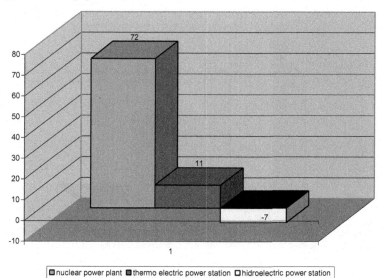

Picture 3. Russian public representation about quantity of the electric power made by atomic power station

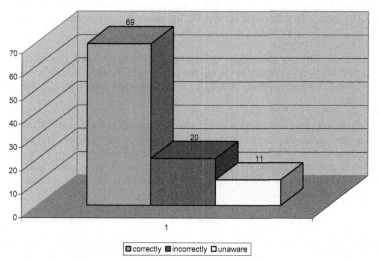

PICTURE 4. AN INDEX OF POWER STATION'S DANGER FOR ENVIRONMENT
(OPINION OF THE EXPERTS).

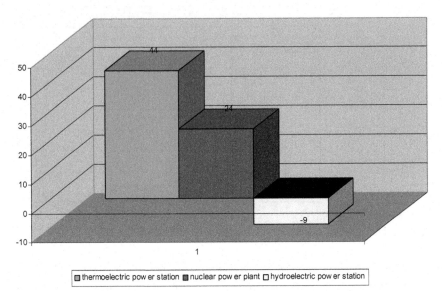

The correlation between the data of interrogation of the teachers and population is observed. In
this connection it is necessary to carry out educational work with the teachers and students.
Unfortunately, there is no understanding that, the formation of public opinion is a long multilevel
work.

Without constant work on formation of representation about nuclear power is one of the
components of steady development to change the existing mainly negative public opinion to this
problem. It will be a constant brake, both with the decision of this question, and with realization
of strategic tasks.

The lack of the objective information about plus and minuses of the use NPS derivates and
continues to derivate the population myths about the threat of the radiation. The problems of an
environment frequently are used by politics in the advertising purposes, thus the scientific facts
are consciously deformed. Basing on opinion of a public which is taking place under influence of
mass media a community in which risks to make an incorrect choice of a spectrum possible
strategic long-term decisions.

The propagation, myths and the fears play a negative role for development of nuclear power
generation. The vivid example to that can be served by a refusal of Germany from construction
and operation NPS in connection with pre-election tactics of the political forces made the rate on
the negative relation of the population of the country to atomic engineering after failure on

Chernobil NPS. The government of France adheres to other tactics in power area in this connection, at present NPS makes a basis of power of the country.

PICTURE 5. AN INDEX OF POWER STATION'S DANGER FOR ENVIRONMENT (OPINION OF THE TEACHERS).

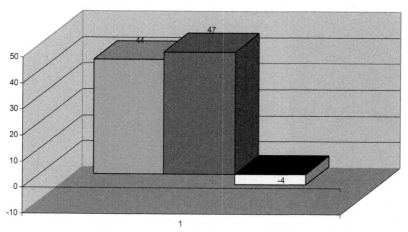

In this connection the unique output for a community, gets rid of the unjustified fear before radiation and manipulations from the party and political forces it is the objective information on advantages and lacks nuclear, thermal and atomic engineering, and this information should be objective and popular. And only on the basis of this information can build the community long-term strategy of steady development of power.

CONCLUSIONS:

Public opinion should be considered before making any long term planning for future power generation. The population should be educated about all the energy sources available to us and how they can be used. the atomic engineering is competitive in comparison with traditional organic fuels on economic parameters and ecological facts. A degree of risk for the personnel engaged in NPS is less than in TPS.

REFERENCES:
1. Risk Professional Mortality. (1996). Social Advantages of Nuclear Power.
2. Radiation - Hygienic Passport of Russian Federation. (1999). Ministry Public Health of Russia.

ANALYSIS OF RADIO ECOLOGICAL SITUATION IN AZERBAIJAN

L. A. ALIYEV
Institute of Radiation Matters, National Academy of Sciences
Ministry of Ecology and Natural Resources
Baku, AZERBAIJAN.

Corresponding author: monitoring@mmd.baku.az

ABSTRACT:

In this article, factors forming radio-ecological state on the territory of Azerbaijan Republic are investigated on the basis of appropriate literature and long-term researches. Along with stated, natural radiation background of some regions within the Republic for 1999-2002 has been studied, radiation of atmospheric precipitations in big industrial cities has been specified for 1993-2001 and radio-spectrometric analysis of radio-ecological state at less polluted Apsheron peninsula was carried out.

Keywords: Azerbaijan Republic, radiation, radio-spectrometric analysis, radionuclides

INTRODUCTION:

Earth being an environment for organic world receives radiation rays from various sources. Most part of it is received by the living beings from radiation sources inherent to earth itself. Depending on its location and constituent rock radionuclide components the radiation level varies in different areas. It is known that most part of radioactive emanation [5,6] is created in earth sources. Radionuclides such as K-40, Rb-87 and U-238, Th-232 usually encountered in mountain rock areas of the earth are members of radioactive family originated from isotopes of half-deterioration period [1]. The rest part of natural emanation falls on external radiation creating cosmic rays. In some cases cosmic rays reaching the earth surface create direct radiation; in other cases cosmic rays being in mutual interference with atmosphere procreate various radionuclides and plays role of second emanation source [3].

For Azerbaijan has variegated soil-climatic conditions radiation background created by both cosmic emanation and rocks, it causes originating of different radiation states on the territory of the country [1]. Artificial factors influencing the radiation state are also inherent in the area of our country. As Azerbaijan is not a nuclear country in comparison with natural emanation, artificial radioactive substances, being a product of human labor, do not play leading role in the formation of radiation state.

Not withstanding there exist some objects to influence or disposed to influence the radiation state in the countries surroundings. The list of such objects may include Armenia Nuclear Power Plant

M.K. Zaidi and I. Mustafaev (eds.), Radiation Safety Problems in the Caspian Region, 241-245.
© *2004 Kluwer Academic Publishers. Printed in the Netherlands.*

(NPP), Rostov NPP, Kazakhstan nuclear water purification unit, Chernobil NPP, nuclear objects in Iran and Iraq, nuclear weapon tests conducted in Middle and South Asia [2]. Among these objects role of Metsamor NPP in Armenia and Chernobil accident (catastrophe) is peculiar. In the process of Metsamor NPP exploitation, the pollution of small rivers flowing into Kura and Araz by the toxic and nuclear waste sites occurs. In addition, clay waters evolved in the process of production of petroleum products originate radioactive man-caused pollution sources on the area of Azerbaijan.

We have been investigating radio-ecological state within the Republic over the years. In spite of that fact, the research conducted is not enough. Systematic research works on the investigation of Sr-90 and Cs-137 radioisotopes have been carried out [1]. It was found that the main ingredients in the enhancement of natural radiation background within Azerbaijan are radionuclides incoming into the environment.

In present work interaction of natural radiation background and radioactivity of atmosphere precipitations on the territory of Azerbaijan is studied. Within 1993-2002 the dynamics of these parameters in number of Azerbaijan regions were also investigated. Radio-ecological states on Apsheron peninsula were illustrated in present work, and radio-spectrometric analysis of polluted areas has been carried out.

EXPERIMENTAL:

Exposition dose intensity of radiation was defined by DRQ-01Q dosimeter. Total β-radioactivity of atmosphere precipitation samples was measured by ascertainment of total β-radioactivity RUB-01P radiometer. Special radioactivity of samples were brought to required level by burning them in muffle furnace. For these purposes vertical plane tables were taken and aerosol deposition gauze parts were heated in close containers to 450^0C, then total β-radioactivity - A_n of samples was calculated in accordance with the following formula:

$$A_\sigma = \frac{N_{C+\sigma} \; \square \; N_\sigma}{60\sigma} \frac{P_\sigma}{P_C} K$$

where K- breaking up reduction, $N_{c+\phi}$ - computation speed of sample together with background (imp/min), N_ϕ - computation speed of background, P_n/P_c- relativity of total weight of sediment obtained at sample combustion to the weight of prepared sediment (K = 1 is taken for background measurement unit). All obtained sediment in usage equals $P_n/P_c = 1$. Percentage of incorrect statistic calculations was implemented by the following formula:

$$\sigma = \frac{\sqrt{N_{C+\sigma}/t_{C+\sigma} + N_\sigma/t_\sigma}}{N_{C+\sigma} \; \square \; N_\sigma} \square 100\%$$

Where $t_{c+\phi}$ and t_ϕ - calculating interval of sample and background computation speeds. Quantity of radionuclides in samples and their identification was specified by means of ORTEX radio-spectrometer.

DISCUSSION:

Ascertainment of natural radiation background in different regions of Azerbaijan.

In the first part of present work natural radiation background in different regions of Azerbaijan were defined. The result of the research work undertaken illustrated in four different diagrams. Analysis of the acquired results for all years shows that natural radiation background on the Republic is in 6-28 µR/hour intervals.

Bilasuvar is a region of relatively intensive natural radiation background, and Zagatala-Qabala are regions of relatively low radiation background. Evidently, that annual results each month in general do not vary much. In case of small changes in some regions are still available, generally upward and down trend of radiation background is not observed.

Zagatala-Qabala zone is characterized by high level of natural radiation background that is more likely to constitute superiority factor of earth stratum among the natural background forming factors in this area.

The results of 1999 show that annual radiation on the territory of Oguz region is decreased with low alternations and such alternations are impossible to connect with processes to take place in earth stratum. Apparently its cause may be explained by cosmic emanation established by direct solar activity on this area.

Such alternations for 2000 were registered in Gandja region. It is interesting that noticeable changes took place in July and August. It is known that in these months solar rays fall on the territory at right angle, and it may cause changes. For 2001 and 2002 as for previous years season regularities were observed in small alternations of natural radiation background.

Ascertainment of total β-radio activity of atmospheric aerosols in different industrial areas of Azerbaijan.

Second part of presented work contains the ascertainment of atmospheric precipitation radioactivity within the borders of different big industrial cities of the Azerbaijan. This increase is more likely to be the direct result of global atmospheric displacement in different parts on the earth, with Republic area included, caused at atmospheric precipitations. In 1994 relatively high radioactivity observed in atmospheric precipitations draws special attention. This year, the measurement results in all big industrial cities atmospheric precipitation radioactivity in comparison with different cities increased for 3-4 times. In mentioned year there can be various reasons for increase of atmospheric precipitation radioactivity. Obviously the process on the territory cannot be the reason of it, thus as we don't possess any nuclear weapon and nuclear tests are not being undertaken in Azerbaijan. The existing anomaly probably is on the account of global atmospheric precipitations. In 1994 in Middle East (India and Pakistan) second wave of radionuclide migration as a result of nuclear weapon tests and Chernobil catastrophe in 1986 and being spread to the surrounding area proves this fact.

Study of radio-ecological state of oil-producing areas on Apsheron peninsula.

In the third part of presented work radio-ecological state of oil-producing areas on Apsheron peninsula is studied. The results of long-term studies show that radio-ecological state on Apsheron peninsula is not satisfactory enough. Local pollution zones differentiating the radio-activity in oil-production areas from natural backgroung in 100-150 times are available. Balakhani Oil and Gas Production Department (OGPD) is mostly characterized by these features. It can be said that this area is completely polluted with well water sediments and zones polluted with 150-2000 μR/hour radioactivity are emerged around some oil boreholes.

Gum Island and area of Azizbayov OGPD also has zones polluted with various radionuclides of 1000 μR/hour intensity emerged. Pollution zones were also registered in oil-processing plants and some settlements.

For instance, in the result of well pipe cleaning OGPD in the area near worker's town of Gum Island were polluted by yellow crumb of 200-800 μR/hour radioactivity. Such local pollution zones can be observed in various workshops at Sattarhan and Surakhani plants.

Investigation of Baku Iodine plant environment revealed that plant area is characterized by 200 μR/hour radioactivity, and artificially emerged ponds, their water is used for production, are characterized by 60-70 μR/hour radioactivity.

Main part in radionuclide pollution of Apsheron peninsula is formed by well water emerged from ponds. Such ponds are located in Log-Batan, 20th area, Bayil, and Amirdjan settlements. In order to discover radionuclide distribution on Apsheron peninsula, the radio-spectrometric analysis of soil and water samples were made.

Analysis revealed the availability of strontium-90, caesium-137, caesium-134, europium–155, europium-154, cerium-144, americium-241, ruthenium-106, stibiate-125, silver-110, thorium-232, uranium-238 and potassium-40 isotopes in these samples. It was registered that:

1. Ce-137, Eu-155, U-238, Th-232, K-40 radio-isotopes are in larger quantity in comparison with other mentioned isotopes.

2. The content of Cs-137 isotope in sandy soils comparatively lower from its content in clay soils, then Eu-156, U-238, Th-232, K-40 isotopes contents approximately twice higher in clay soils.

3. The content of Sr-90 isotope in top layer of clay soil (0.5Bk/kg) is three times higher from sand soil (0-5 cm). On getting into depth (0-25 cm) the quantity of this isotope is slumping in clay soil, and has fuzzy decline in sandy soil.

So it becomes clear that in clay soil, Sr-90 isotope is disposed to sorbtion by its top layer, and in sandy soils it is equally distributed at 0-25 cm depth. Radio-spectrometric analysis shows that oil samples polluting environment of wells contain uranium, radium and thorium isotopes. The appropriate range of isotope volume in these samples is in 2, $3-13.6 \times 10^{-7}$ g/kg, $1.2-3.1 \times 10^{-12}$ g/kg and $8.2-17.0 \times 10^{-7}$ g/kg interval. In highly radioactive ponds there are uranium, radium and thorium in $1.2-4.1 \times 10^{-6}$ g/l, $2.5-3.2 \times 10^{-11}$ g/l and $1.1-3.1 \times 10^{-7}$ g/l volume.

CONCLUSION:

According to results obtained, it was concluded that artificial radionuclides cesium-137 and strontium-90, as a result of global atmosphere precipitations, are available in Apsheron soils. Natural radionuclides potassium-40, uranium-238 and thorium-232 are microelements specific for Apsheron soil. Potassium–40 with great likelihood is a product of salinity, and Uranium-238 and Thorium-232 of well waters.

REFERENCES:

1.   Aliyev D.A, Abdullayev M.A. (1983). Strontium-90 and cesium-137 in Soil-vegetable Cover of Azerbaijan. Nauka, Moscow.
2.   Belousova I.M, Shtikkenberg Y.M. (1961). Natural Radioactivity, State Publishing House of Medical Literature. Medgiz, Moscow.
3.   Translation from English. Bannikov Y.A. (1988). Radiation. Doses, Effects, Risk. Publishing House "Mir". Moscow.
4.   Makhonko K.P. (1982). Survey on Radioactive Pollutions of Natural Environment. Leningrad.
5.   Aliyev L.A. (2002). Radiation Protection Systems in Azerbaijan. INIS. Vienna.
6.   Jafarov E.S. (2003). The Ecological Problems Created in the Absheron Peninsula by the Oil Extraction and Production, Baku.

SUBJECT INDEX

^{137}Cs, 1, 5, 6, 7, 8, 9, 10, 11, 51, 52, 76, 77, 106, 179, 180, 182, 183, 226, 227

^{241}Am-pollution, 51

analysis, 1, 3, 4, 5, 13, 19, 20, 21, 23, 30, 33, 44, 57, 58, 59, 61, 66, 72, 75, 76, 77, 82, 98, 99, 113, 122, 149, 153, 158, 160, 167, 185, 191, 194, 198, 199, 213, 232, 234, 244

Aral Sea basin, 13, 15

automated monitoring, 1

Azerbaijan, vi, v, vi, vii, viii, x, xii, xiii, xiv, xv, xix, xx, xxi, xxii, 1, 2, 3, 4, 17, 18, 19, 23, 57, 84, 95, 97, 98, 99, 102, 103, 104, 106, 119, 121, 130, 133, 134, 135, 141, 150, 151, 152, 157, 162, 163, 165, 166, 167, 168, 170, 175, 176, 194, 203, 205, 209, 216, 217, 219, 222, 223, 241, 242, 243, 245

Azerbaijan Republic, 1, 4, 150, 157, 175, 176, 241

Azgir Nuclear Test Site, 29

background, 1, 2, 3, 4, 17, 18, 21, 30, 33, 34, 41, 70, 73, 81, 83, 99, 104, 115, 133, 147, 148, 149, 150, 157, 158, 159, 160, 177, 181, 242, 243

bioelectromagnetics, 211

biological action, 217

biotopes, 97, 101

border monitoring, 169, 171, 172

borosilicate, 23, 24, 25, 26

cancer, 107, 108, 110, 111, 112, 113, 114, 115, 187, 220, 223

chromosome aberrations, 51, 52, 53, 55, 56

coal, 80, 131, 144, 231, 232, 233, 234, 235

compatibility, 217, 218

complex, 3, 15, 16, 18, 24, 56, 79, 85, 86, 87, 89, 90, 108, 118, 121, 122, 124, 127, 146, 148, 152, 173, 175, 211, 212, 215, 223

composite material, 205, 206, 207, 208

conformation, 57, 58, 59, 61, 62, 65, 66, 128

contamination, 5, 11, 13, 29, 30, 32, 70, 73, 75, 78, 79, 98, 101, 102, 113, 117, 118, 129, 131, 174, 185, 191, 227

control, 2, 3, 15, 16, 51, 52, 53, 69, 78, 90, 108, 110, 111, 117, 125, 127, 131, 133, 157, 167, 169, 170, 171, 172, 185, 188, 189, 214, 217, 221, 235

cooperation, 7, 161, 162, 225, 228

data collection, 5

desertification, 13, 14, 15

developing transit countries, 165

dimethyl sulfoxide, 121

dirty bomb, 165

early warning, 33

ecolimnologic, 97

ecological catastrophes, 1, 2

ecology, 2, 13, 79, 147, 173, 175, 176, 191, 205, 218

electret, 205, 206, 207

electromagnetic field, 212, 214, 217, 218, 220, 221, 222

emergencies, 1, 2, 118

EMF, 211, 218, 220, 221, 222

environment, 1, 2, 3, 4, 5, 7, 11, 15, 17, 18, 19, 33, 39, 57, 59, 69, 70, 78, 79, 81, 97, 99, 102, 109, 113, 130, 134, 147, 148, 151, 157, 160, 161, 173, 174, 175, 176, 191, 205, 212, 217, 219, 225, 227, 228, 231, 234, 235, 237, 238, 239, 241, 242, 244

environmental fields, 211

export control, 165, 169, 170, 172

exposure, 5, 29, 69, 70, 107, 108, 109, 110, 113, 115, 120, 129, 130, 131, 132, 133, 134, 147, 148, 149, 150, 155, 177, 182, 186, 187, 188, 197, 211, 212, 213, 214, 215, 216

exposure dose rate, 70, 147, 149, 150

extraction, 43, 44, 45, 46, 47, 48, 50, 79, 89, 90, 91, 92, 93, 95, 130, 134, 135, 141, 148, 151, 174, 203

fallout, 6, 10, 69, 76, 227

fertilizer, 18, 43, 89, 197, 198

food, 10, 11, 16, 132, 161, 197, 227
four-dimensional ions, 23
fractions, 29, 31, 32, 113, 141, 151, 152
gamma radiation, 13, 16, 113, 114, 133, 197
gamma survey, 70, 157, 160
gamma-irradiation, 51, 53, 56
gas, 18, 129, 130, 131, 132, 133, 134, 136, 137, 138, 139, 142, 143, 145, 147, 149, 150, 151, 152, 153, 154, 155, 160, 191, 192, 193, 194, 202, 203, 231, 234, 235
gases, 1, 4, 131, 136, 137, 141, 142, 143, 144, 145, 152, 154, 191, 205, 235
genotoxicity, vi, 51, 56
granulometrical fractions, 29, 32
heavy metals, 15, 17, 18, 19, 20, 21, 79, 155, 235
heavy metals in soils of industrial areas, 17
heptane, 124, 135, 136, 137
human health, 191, 213, 223
hydro-electric, 231
illicit trafficking, 165, 166
instrumentation, 3, 217
irrigation, 13
kerosene, 43, 44, 45, 46, 47, 48, 50, 90, 151, 153, 154, 155
kerosene oil, 151
liquid products, 138, 141
measurement, 3, 4, 30, 34, 35, 36, 152, 158, 160, 177, 179, 180, 187, 198, 206, 215, 216, 217, 219, 221, 222, 242, 243
medical dosimetry, vii, 177, 181, 184, 189
medical radiation, 107, 117
medicine, 87, 103, 108, 110, 111, 112, 115, 121, 122, 127, 161, 166, 186, 189
mineralization, 13, 14, 73, 74, 102, 148
mining, 18, 69, 85, 87, 131, 166
molecular dynamics simulation, 212
monitor, 1, 172, 217
monitoring, xii, 2, 3, 4, 11, 16, 17, 33, 70, 75, 76, 84, 109, 112, 119, 158, 169, 171, 178, 180, 185, 186, 205, 217, 220, 241
morphometric, 97
natural gas, 130, 151, 152, 153, 154, 155, 232
nonproliferation, 161, 165, 171
nuclear export control, 165

nuclear material, 1, 165, 166, 167, 173, 174, 175
nuclear materials, 1, 165, 166, 167, 174, 175
nuclear power, 85, 174, 175, 227, 228, 231, 232, 234, 235, 236, 238
nuclear security, 166, 167, 168
nuclear terrorism, 165, 166, 167, 173, 175, 176
nuclear weapons, 6, 161, 165, 166, 173, 174, 175
oil, 18, 21, 97, 98, 99, 100, 101, 102, 123, 129, 130, 131, 132, 133, 134, 135, 138, 139, 141, 144, 145, 146, 147, 148, 149, 150, 151, 152, 153, 154, 155, 157, 160, 166, 235, 244
oil-bituminous rocks, 145, 146
personal dosimetry, 177, 178, 179, 180, 181, 182, 183, 184, 185, 188
petroleum, 18, 231, 232, 234, 235, 242
phosphoric acid, 43, 44, 45, 46, 47, 50, 89, 90, 91, 95, 197, 200, 202
polyene antibiotics, 121, 123
polymer thin film, 205
population, 5, 7, 10, 11, 13, 14, 16, 56, 69, 75, 78, 81, 82, 147, 165, 186, 188, 217, 219, 221, 225, 227, 228, 231, 233, 234, 235, 236, 237, 238, 239
protection, 12, 41, 75, 108, 134, 147, 155, 165, 171, 173, 175, 176, 177, 180, 185, 187, 188, 221, 223, 225, 228, 234
purification, 10, 135, 139, 242
radiation, v, 1, 2, 3, 4, 5, 7, 10, 11, 12, 16, 23, 25, 27, 29, 31, 33, 34, 35, 37, 38, 41, 51, 56, 70, 77, 83, 84, 88, 99, 107, 108, 109, 110, 111, 112, 113, 114, 115, 116, 117, 118, 119, 120, 125, 129, 131, 132, 133, 134, 135, 136, 137, 138, 139, 141, 142, 143, 144, 145, 147, 148, 149, 150, 152, 155, 157, 158, 159, 160, 167, 177, 178, 179, 183, 184, 185, 186, 187, 188, 189, 194, 205, 206, 220, 225, 228, 234, 238, 239, 241, 242, 243
radiation background, 29, 70, 147, 148, 149, 150, 157, 160, 241, 242, 243

radiation dose, 34, 35, 107, 109, 111, 112, 113, 177, 178, 179, 187, 188, 189, 205, 206
radiation map, 3, 35, 157
radiation monitor, 1, 5, 16, 33, 157
radiation protection, 33, 107, 108, 118, 129, 133, 134, 177, 184, 186, 188
radiation safety, 1, 2, 4, 10, 16, 107, 118, 167, 225, 228
radioactive materials, 80, 88, 165, 167, 174
radioactive sources, 114, 119, 165, 167
radioactive waste, 2, 29, 69, 70, 75, 78, 118, 134, 147
radioactivity, 17, 18, 83, 97, 98, 99, 101, 102, 103, 113, 115, 129, 133, 147, 148, 149, 153, 155, 197, 198, 199, 200, 202, 235, 242, 243, 244
Radioactivity, vi, vii, 22, 97, 99, 152, 155, 197, 204, 229, 245
radiological safety, 79
radiolysis, 27, 135, 138, 144
radionuclide, 3, 29, 32, 51, 52, 70, 72, 76, 130, 160, 202, 241, 243, 244
radionuclides, 1, 3, 5, 6, 7, 8, 9, 11, 18, 29, 31, 51, 70, 72, 73, 76, 89, 97, 98, 99, 101, 102, 103, 104, 105, 106, 111, 113, 129, 131, 132, 134, 147, 148, 151, 152, 160, 197, 198, 200, 202, 227, 235, 241, 242, 243, 244, 245
radioprotector, 121
radio-spectrometric analysis, 241, 242, 244
radium, 83, 97, 99, 103, 129, 131, 132, 133, 149, 160, 197, 202, 203, 244
radon, 70, 75, 97, 99, 129, 130, 132, 133, 151, 152, 153, 154, 155, 197, 198, 202, 203
RESA, v, 33, 36, 41
research, xxii, 3, 7, 15, 18, 19, 80, 82, 83, 85, 87, 89, 103, 106, 108, 111, 148, 149, 160, 161, 167, 169, 173, 180, 185, 191,

198, 203, 211, 213, 214, 216, 217, 220, 222, 223, 225, 226, 228, 229, 235, 236, 242, 243
RF safety, 211
risk assessment, 5, 129, 134, 189
safety, 1, 2, 12, 16, 41, 79, 112, 115, 118, 119, 134, 170, 175, 186, 189, 211, 212, 217, 218, 219, 222, 223, 236
soil pollution, 17, 52
sources of soil pollution, 17
support, xxii, 7, 12, 84, 95, 119, 125, 134, 162, 163, 174, 189, 203, 206, 211, 213, 216, 222, 225, 228
survey, 5, 7, 19, 30, 70, 71, 111, 116, 216
terrorism, 165, 173, 174, 175, 176
thermal power plants, 231
thorium, 79, 83, 85, 87, 89, 97, 99, 101, 103, 104, 158, 159, 160, 202, 244, 245
three-dimensional, 23, 25
tobacco, 103, 106
TOPO, 89, 90, 91, 92, 93, 95
Tradescantia, 51, 52, 53, 56
typological character, 97, 98
tyrosine hydroxylase, vi, 57, 66, 67
underwater gamma spectrometer, 157
uranium, 43, 44, 45, 46, 47, 48, 49, 50, 69, 79, 81, 83, 87, 89, 90, 91, 92, 93, 94, 95, 97, 99, 101, 103, 158, 159, 160, 166, 175, 199, 203, 233, 235, 244, 245
water, 1, 3, 10, 11, 13, 14, 15, 18, 19, 23, 29, 32, 50, 70, 73, 74, 75, 76, 77, 78, 81, 84, 90, 91, 97, 98, 99, 102, 122, 123, 124, 125, 127, 128, 131, 132, 135, 136, 137, 138, 139, 141, 142, 143, 144, 147, 148, 157, 159, 174, 191, 192, 198, 232, 236, 242, 244
weapons, 6, 169, 171, 174, 175
wet process, 43, 89, 90, 91, 95
x-rays, 107, 109, 110, 112, 113, 114